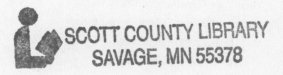

The Mad Scientist Handbook 2

The Mad Scientist Handbook 2

How to Make Your Own

Disappearing Ink,

Exploding Egg,

Smoking Fingertips,

Flying Potato,

Rain Machine,

X-Ray Glasses,

and Much More!

Joey Green

A Perigee Book

Warning: A responsible adult should supervise any young reader who conducts the experiments in this book to avoid potential dangers and injuries. The author has conducted every experiment in this book and has made every reasonable effort to ensure that the experiments are safe when conducted as instructed; however, neither the author nor the publisher assume any liability for damages caused or injury sustained from conducting the projects in this book.

A Perigee Book
Published by The Berkley Publishing Group
A division of Penguin Putnam Inc.
375 Hudson Street
New York, New York 10014

First Perigee edition: July 2002

Visit our website at www.penguinputnam.com

Library of Congress Cataloging-in-Publication Data

Green, Joey.
    The mad scientist handbook 2 / by Joey Green.
        p. cm.
    Includes bibliographic references.
    Summary: A collection of experiments that can be
performed using ordinary household objects, such as
making a battery from money or a beach ball–powered
elevator, plus explanations of why each works.
    ISBN: 0-399-52775-3
    1. Science—Experiments—Juvenile literature. [1.
Science —Experiments. 2. Experiments.] I. Title: Mad
scientist handbook two. II. Title.

Q164 .G732 2002
793.8—dc21
                                                2001058025

Printed in the United States of America

10   9   8   7   6   5   4   3   2   1

For
Ashley, Julia,
Eric, Rebecca, Andrew, Lauren,
Matthew, Sammy,
Jonathan, Alexander, Zachary,
Corinne, Janelle,
Kailey, Shannon, and Devin

Contents

Introduction

I love making a mess and creating freaky things. That's what being a Mad Scientist is all about. I'm not interested in teaching you anything other than how to have a blast making Monster Bubbles, Alka-Seltzer–Powered Rockets, and Exploding Coca-Cola—all by simply mixing up stuff you probably have around your house right now. If you happen to learn something along the way (like chemistry and physics) and wind up getting a Ph.D. at MIT in some weird subject like rocket science, well, don't blame me. I can't help it if your kitchen just happens to double as a chemistry lab-

oratory. It's not my fault that someone filled your pantry with chemicals like acetic acid (better known as Heinz vinegar), citric acid (ReaLemon lemon juice), sodium bicarbonate (Arm & Hammer baking soda), aluminum (Reynold's Wrap), and potassium bitartrate (cream of tartar). I'm just here to show you how to make really cool slimy-gooey things and wild doohickies and awesome thingamabobs.

Okay, so I just happened to be the kid who won the science fair back in the sixth grade. It wasn't my fault. It was just an accident. Really. During the summer of the *Apollo 11* mission to the moon, I decided it would be fun to cut out every article that appeared in my hometown newspaper about *Apollo 11* and glue them into an enormous notebook. Naturally, I cut up the newspaper before my father got home from work to read it—which was half the fun and nearly provoked my dad to send me to the moon without a space suit, rocket or Tang drink mix. Enamored with the space program, I also bought a model kit of a three-foot-tall replica of the *Apollo 11* rocket and, although I was only eleven years old, I somehow found the patience to follow the directions to the letter. The completed rocket sat on my desk, much to the glee of my younger brother Doug, who constantly knocked it over. Or maybe that was my dog, Smokey.

When the sixth-grade science fair rolled around and I had no idea what to do as a project, my mother wisely urged me to bring in my two over-stuffed notebooks of yellowing newspaper clippings and my abused model of the *Apollo 11* rocket. Being a kid who hated school (and homework in particular), I jumped at my mother's brilliant suggestion. In school, I used my model to demonstrate the various stages of the *Apollo 11* mission—how the command module docked with the lunar module in space, how the lunar module blasted off from the moon's surface, and how only the command module returned to earth.

When I was selected winner of the science fair for the entire sixth grade, I was shocked. I thought I had been rewarded for my laziness. In truth, I had been rewarded for my passion outside the classroom. A little encouragement can be a dangerous thing. Even now, as my sixth-grade science fair trophy collects dust on a shelf, it inspires me to constantly explore strange new worlds—by mixing shaving cream with glue to make foam, making giant bubbles and racing Slinkys down staircases.

Welcome to my world. I hope you enjoy your stay.

Alka-Seltzer-Powered Rocket

WHAT YOU NEED

- ☐ Fuji 35mm film canister
- ☐ Scissors
- ☐ Scotch tape
- ☐ Sheet of colored construction paper
- ☐ 1 cup cold water
- ☐ Alka-Seltzer tablets

WHAT TO DO

On Fuji 35mm film canisters, the lid snaps inside the body. Roll the sheet of construction paper around the film canister, so the open end of the film canister sticks out. Tape the paper onto the film canister. Using scissors and Scotch tape, add a paper cone to the top of the paper tube.

Turn the rocket upside-down so the open end of the film canister faces upward. Fill the canister halfway with cold tap water. Drop in two Alka-Seltzer tablets. Snap on the lid, turn the rocket rightside-up, set it down on the ground, and quickly step back.

WHAT HAPPENS

The rocket blasts approximately six feet off the ground.

WHY IT WORKS

When activated in water, the Alka-Seltzer tablet releases carbon dioxide gas, filling the sealed canister—until the pressure becomes too great, popping the canister off its lid. The pressure blows the gas out of the canister, pushing the canister with an equal force in the opposite direction. As Sir Isaac Newton's third law of motion states: For every action, there is an equal and opposite reaction.

BIZARRE FACTS

- Alka-Seltzer is a coined word that suggests *alka*linity and the carbonation of *seltzer*.
- In 1928, Hub Beardsley, president of Dr. Miles Laboratories, discovered that the editor of the local newspaper in Elkhart, Indiana, prevented his staff from getting influenza during a severe flu epidemic by giving them a novel combination of aspirin and baking soda. Beardsley immediately set his chief chemist, Maurice Treneer, to work devising a tablet containing the two ingredients.
- An Alka-Seltzer tablet fizzing in a glass of water prompted a hungover W. C. Fields to joke, "Can't anyone do something about that racket?"

- The original six-inch-high Speedy Alka-Seltzer working model became so famous that it was insured for $100,000 and kept in the vault of a Beverly Hills bank.
- In 1955, a plastic Speedy Alka-Seltzer doll was issued in a limited edition.
- The buffered aspirin in Alka-Seltzer peaks within thirty minutes, whereas regular aspirin tablets peak in about two hours.
- In the 1970s Alka-Seltzer became widely known for its innovative television commercials, launching the catchphrases "Momma mia,

SPEEDY ALKA-SELTZER

Early promotions for Alka-Seltzer featured Speedy Alka-Seltzer, a baby-faced puppet with red hair and a tablet-shaped hat created in 1951. Stop-motion animation brought Speedy to life in 212 television commercials between 1954 and 1964, requiring nineteen plaster heads with various lip positions, two sets of legs and arms, and as many as 1,440 adjustments for a single sixty-second commercial. Speedy Alka-Seltzer co-starred with Buster Keaton, Martha Tilton, Sammy Davis, Jr., and the Flintstones.

that's a spicy meatball," "Try it, you'll like it" and "I can't believe I ate the whole thing."

- The "Plop, Plop, Fizz, Fizz, Oh What a Relief It Is!" vintage theme song for Alka-Seltzer, written by Tom Dawes in 1977, remains one of the most recognized commercial melodies and a favorite of popular culture trivia buffs.

- The only man-made structure visible from space is the Great Wall of China.

- In space, astronauts are unable to cry, because there is no gravity and the tears won't flow.

- If the Spaceship Earth geodesic sphere at EPCOT were a golf ball, to be the proportional size to hit it, you'd be two miles tall.

Battery Madness

Money Battery

WHAT YOU NEED

- ☐ Quarter
- ☐ Coffee filters
- ☐ Pencil
- ☐ Scissors
- ☐ Twelve pennies (dated after 1983)
- ☐ Twelve zinc washers
- ☐ Two feet electrical wire (22 gauge)
- ☐ Wire cutters
- ☐ Nail
- ☐ Electrical tape
- ☐ Bowl
- ☐ ReaLemon lemon juice
- ☐ Compass

WHAT TO DO

Place the quarter on a coffee filter and trace around it with the pencil to create a circle. Repeat this eleven times, creating a total of twelve circles. With the scissors cut out the twelve circles.

Place a penny on the tabletop. Place a circle of coffee filter on top of the penny. Place a zinc washer on top of the blotter paper. Place another penny on top of the zinc washer, followed by

securing all the coins together (as if using a rubber band).

Fill the bowl with ReaLemon lemon juice. Drop in the stack of coins. Move the head of the nail over the point of the compass needle.

a second circle of coffee filter, followed by another zinc washer. Repeat until you have stacked all the pennies, circles of coffee filter, and zinc washers—ending with a zinc washer on top.

With adult supervision, use the wire cutter to strip one inch of plastic coating off each end of the wire. Wrap the middle of the wire around the nail tightly until the nail is covered with wire. Peel off a four-inch strip of electrical tape. With the scissors, carefully cut the strip of tape down the center to make two narrow four-inch strips of tape. Using one strip of the tape, attach the end of one wire to the top coin and the end of the second wire to the bottom coin, while simultaneously

WHAT HAPPENS

The coffee filter disks, saturated with lemon juice and sandwiched between two coins made from different metals, creates a wet cell battery with enough voltage to make an electromagnet with the strength to move a compass needle.

WHY IT WORKS

An electrolyte (the citric acid in the lemon juice) between two electrodes of different chemically active material causes one of them, called an anode (the copper in the pennies), to become negatively charged, and the other, called a cathode (the zinc in the washers), to become positively charged.

Pot Scrubber Battery

WHAT YOU NEED

- ☐ Two feet electrical wire (22 gauge)
- ☐ Wire cutters
- ☐ Nail
- ☐ Copper scouring pad
- ☐ Paper towels
- ☐ White vinegar
- ☐ Aluminum foil
- ☐ Compass

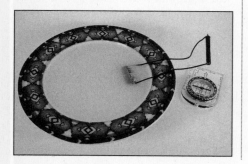

WHY IT WORKS

The vinegar (containing acetic acid) is the electrolyte, the copper scouring pad is the anode, and the aluminum foil is the cathode, creating a wet cell battery with enough voltage to make an electromagnet with the strength to move a compass needle.

WHAT TO DO

With adult supervision, use the wire cutters to strip one inch from both ends of the wire. Wrap the middle of the wire around the nail tightly until the nail is covered with wire.

Secure one end of the wire to the cooper scouring pad. Saturate a sheet of paper towel with vinegar and then wrap it tightly around the copper pad. Wrap a sheet of aluminum foil tightly around the paper towel. Attach the other end of the wire to the aluminum foil.

BIZARRE FACTS

■ In the 1790s, Italian scientist Count Alessandro Volta made the first battery. The volt, a unit of electric measurement, is named after him. Volta's first battery consisted of stacked pairs of silver and zinc disks separated from one another by cardboard disks moistened with a salt solution.

■ Instead of using lemon juice as the electrolyte, you can use vinegar (acetic acid) or salt water (sodium chloride).

■ Most pennies minted before 1983 are made from a copper-zinc alloy, while pennies made after 1983 are made from zinc coated with a thin layer of copper.

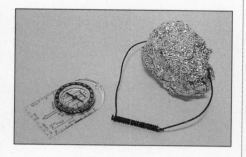

Move the head of the nail over the point of the compass needle.

WHAT HAPPENS

The compass needle moves.

SIT RIGHT BACK AND YOU'LL HEAR A TALE

In an episode of the television comedy series *Gilligan's Island* [Episode 18: "X Marks the Spot"], to recharge the batteries in the castaways' radio, the Professor sets up several metal strips, pennies, and coconut shells filled with seawater, and instructs his fellow castaways to stir the seawater in each coconut shell.

Beach Ball Elevator

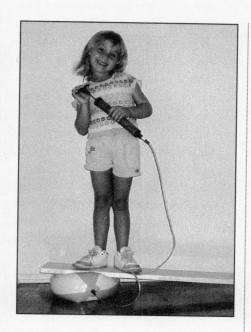

WHAT YOU NEED

- ☐ Four-foot-long vinyl flexible hose (¼ inch in diameter)
- ☐ Beach ball (1 foot in diameter)
- ☐ Electrical tape
- ☐ Scissors
- ☐ Inflating hand pump
- ☐ 1 plank of ¾-inch thick wood (1-by-5 feet)

WHAT TO DO

Insert one end of the flexible hose into the open nozzle of the beach ball and secure in place with a piece of the electrical tape. Insert the nozzle of the basketball hand pump into the other end of the flexible hose and secure in place with another piece of electrical tape. Set the plank of wood on a clean, firm surface (sidewalk, tile floor) and place the uninflated beach ball under one end of the plank of wood. Stand on the other end of the plank to hold it in place. Have someone stand on top of the wood over the beach ball. Pump the basketball pump.

WHAT HAPPENS

As the beach ball fills with air, the plank of wood slowly rises, elevating the person standing on it.

WHY IT WORKS

French scientist and philosopher Blaise Pascal (1632–1662) discovered that a fluid in a container transmits pressures equally in all directions. This principle, known as Pascal's Law, explains how hydraulic lifts work. The air blown into the beach ball distributes pressure equally throughout the ball. If air at ten pounds of pressure per square inch is blown into the ball and ten square inches of the ball touch the piece of wood, the air will lift one hundred pounds of weight.

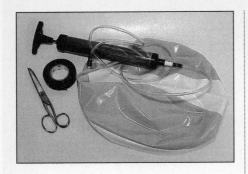

BIZARRE FACTS

- Some ants can lift fifty times their own weight.
- On July 19, 1989, an engine exploded on an airborne United Airlines DC-10, crippling the hydraulic power system and paralyzing all the flight-control surfaces of the plane. Pilot Alfred C. Haynes and his crew managed to crash-land the plane in Sioux City, Iowa, by alternately accelerating and decelerating the right and left engines. The plane caught fire during the landing, killing 112 people, but 184 passengers and crew members survived.
- In 1993, Amtrak's X2000 high-speed train began operating on the Washington–New York Metroliner route, capable of traveling up to 155 miles per hour and using a hydraulic tilting system to take curves forty percent faster.
- Hydraulic lifts allow specially designed city buses and vans to rise and lower to the curb to accommodate physically challenged passengers and those confined to wheelchairs.
- Since the 1960s, the overflow from Lake Okeechobee has been channeled through a massive flood-control project—1,400 miles of canals and hydraulic pumps—to water level in the Everglades in ecological balance.
- The Bastille Opera House in Paris, France, contains a modular concert hall whose height, seating, and proscenium opening can all be altered with hydraulic lifts.
- Poet T. S. Eliot's father was president of the St. Louis Hydraulic-Press Brick Co.

IT'S ALL WATER UNDER THE BRIDGE

The world *hydraulic* stems from the root *hydro*, meaning "water"—despite the fact that hydraulic machines usually use liquids and gases other than water in their workings.

Bottled Clouds

WHAT YOU NEED

- ☐ Clean, empty, 32-fluid-ounce Gatorade bottle
- ☐ Teapot
- ☐ Water
- ☐ 1 ice cube

WHAT TO DO

With adult supervision, fill the teapot with water and bring it to a boil. Carefully fill one-quarter of the empty Gatorade bottle with the boiling water. Place the ice cube across the open mouth of the bottle.

WHAT HAPPENS

A small cloud forms inside the bottle.

HOW IT WORKS

The steam from the water rises, but as it nears the ice cube, the steam begins to cool, and the water vapor in the steam condenses, turning into tiny water droplets, creating a cumulus cloud.

BIZARRE FACTS

- Clouds are simply masses of tiny water droplets floating in the air.
- Since water vapor is heavier than air, all clouds should eventually fall

from the sky. But the droplets of water in clouds are so small and light, the slightest air current keeps them aloft. When clouds become heavier they fall faster, usually as rain.

■ Fear of clouds is called nephophobia. The word *nepho* is Greek for "cloud."

■ *Voyager 2* sent back images of the planet Neptune that showed a small, peculiarly shaped white cloud that travels around Neptune roughly every sixteen hours. Scientists nicknamed the cloud "The Scooter."

■ A ray of sunlight shining through breaks in high clouds and illumi-nating dust particles in the air is called a crepuscular ray.

■ A group of small clouds drifting beneath a bigger cloud from which precipitation is falling are called a scud.

WHY ARE CLOUDS WHITE?

Clouds appear to be white because the many small droplets of water vapor making up the cloud scatter the different wavelengths of light equally, so we see a mixture of all the colors of the spectrum—red, orange, yellow, green, blue, indigo, and violet—which together create white.

Brainy Ball

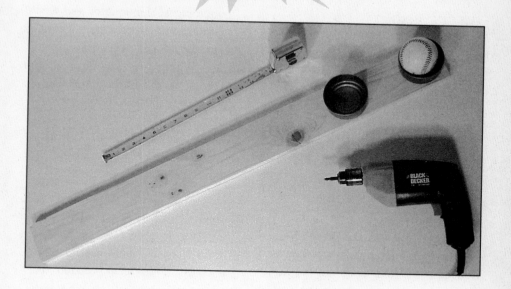

WHAT YOU NEED

- ☐ Electric drill with a Phillip's head screwdriver bit
- ☐ 2 ½-inch wood screws
- ☐ 2 clean, empty six-ounce tuna fish cans
- ☐ 1 plank of ¾-inch thick wood (3-by-36 inches)
- ☐ 1 yardstick or tape measure
- ☐ 1 baseball

WHAT TO DO

With adult supervision, use the drill to screw one tuna fish can at the end of the piece of lumber. Screw the second can to the lumber, three inches away from the first can.

Place the piece of lumber on the floor, perpendicular to a wall, and slide the free end against the wall. Place the baseball in the can nearest the end, lift that end of the lumber thirty inches from the ground (leaving the other end of the ground against the wall). Let the piece of wood drop to the ground.

WHAT HAPPENS

When the lumber falls to the floor, the baseball in the first can falls into the second can.

WHY IT WORKS

Since gravity pulls all falling objects to earth at the same speed, you would think the baseball would stay in the

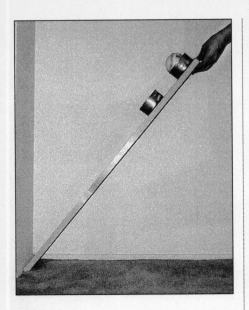

first can. However, the piece of wood is not a free-falling object. The end touching the floor does not fall at all, and it actually pulls down the rest of the wood faster than gravity. The baseball, unattached to the wood or can, falls freely at the constant rate of thirty-two feet per second. The can drops faster than the baseball, the baseball falls straight down, the can (attached to the wood) follows the arc of a circle to land below the falling baseball.

BIZARRE FACTS

- The piece of wood's center of gravity falls at the speed of gravity. When one end of the piece of wood is lifted up, the lumber's center of gravity is approximately twelve inches away from the free end.
- Objects do not weigh the same at all places on the earth's surface.

Because the earth is not perfectly round and it rotates, a person weighing 150 pounds in New York City would weigh 149 pounds standing at the equator and would weigh a little more than 154 pounds standing at the North Pole.

- You weigh less standing on the top of a mountain than you do

THE APPLE DOESN'T FALL FAR FROM THE TREE

Most of Sir Isaac Newton's early biographers fail to mention the story that Newton discovered gravity after watching an apple fall from a tree—casting serious doubt over whether the incident ever really occurred. The two sources of the tale are the French philosopher Voltaire and the Reverend William Stukely, neither of whom witnessed the actual event. Voltaire, in his book *Elements of Newtonian Philosophy*, published in 1738 (eleven years after Newton's death and seventy years after the alleged incident), reported that Newton told the story to his niece, Catherine Barton Conduitt, who cared for Newton in his later years. Reverend Stukely, in his biography of Newton (written in the eighteenth century, but not published until the twentieth century), reported that the physicist told him about the incident while the two were having tea together in the apple orchard at Newton's home.

standing at sea level (because the force of gravity diminishes the farther you are from the center of the earth).

- If you run eastward (in the direction of the earth's spin) you weigh fractionally less than you do standing still. The mild centrifugal force created by the earth's rotation counteracts the effect of gravity. If you run westward (against the earth's spin), you will weigh slightly more. A forty thousand-ton ship sailing east at twenty knots along the equator (where the centrifugal effect is greatest) weighs approximately six tons less than the same ship sailing west.

Butter Machine

WHAT YOU NEED

- ☐ 1 pint of whipping cream
- ☐ Clean, empty, resealable, air-tight plastic container
- ☐ Yellow food coloring
- ☐ Spoon

WHAT TO DO

Pour the entire pint of cream into the plastic container, seal the lid, and shake the sealed container for fifteen minutes or more. Slowly pour out the milky liquid, leaving the butter in the container. Add two drops of yellow food coloring and mix the butter with a spoon. Store the butter in the refrigerator.

WHAT HAPPENS

The cream separates into butter and a milky liquid called buttermilk.

WHY IT WORKS

Milk and cream contain an emulsion of butterfat in the form of tiny, invisible droplets. Shaking or churning causes the butter granules to cling together to form butter.

BIZARRE FACTS

- Ancient Romans used butter as a hairdressing cream and as a skin cream.
- Cream contains roughly ten times more butterfat than milk does.
- Cream can be made by simply allowing a glass of raw milk to stand undisturbed overnight. Gravity causes the milk to sink to the bottom of the glass and the cream to rise to the surface. The layer of cream can be spooned off. The remaining milk is called skimmed milk.
- Wisconsin produces more butter than any other state, followed by California and Minnesota.
- The natural color of butter varies from pale to deep yellow, depending on the breed of cow and what it was fed. Cows eating fresh green grass produce a deep yellow butter while cows eating grain or hay produce a paler color butter. Butter makers usually add food coloring to butter to make it more attractive to consumers.
- Machines cut butter into rectangular sticks called *prints*.
- Butter can be made from the milk of cows, goats, horses, reindeer, sheep, yaks, and other animals.
- In the United States, people use twice as much margarine as butter. Margarine, made from vegetable oil, costs less than butter and contains less cholesterol—a fatty substance that many scientists believe causes arteriosclerosis in human beings, a disease that can lead to a heart attack.
- Butter lasts up to two weeks in the refrigerator. It can be stored frozen at 0° Fahrenheit for up to six months.

BUFFALO BUTTER

People started making butter as early as 2000 B.C.E., when people in India began making butter from the milk of water buffaloes.

Dancing Ping-Pong Ball

WHAT YOU NEED

- ☐ 5-inch-long plastic comb
- ☐ Wool sweater
- ☐ Ping-Pong ball

WHAT TO DO

On a cold day, rub the comb on the wool sweater for one minute. Place the Ping-Pong ball on a smooth, flat surface. Hold the comb vertically an inch above the Ping-Pong ball and slowly spin the comb in circles above the ball.

WHAT HAPPENS

The Ping-Pong ball dances and spins, following the movements of the comb.

HOW IT WORKS

Like charges repel, unlike charges attract. Rubbing the plastic comb on the wool sweater charges the comb with static electricity—transferring electrons from the wool to the comb and giving the comb a negative charge. The Ping-Pong ball remains uncharged. These unlike charges attract each other, drawing the uncharged Ping-Pong ball toward the charged comb.

BIZARRE FACTS

■ If you touch the negatively charged comb to the Ping-Pong ball, the ball will become negatively charged and be attracted to the comb.

- When you comb your hair quickly on a dry day, your hair loses electrons and becomes positively charged, while the comb gains electrons and becomes negatively charged. As you comb your hair, the static electricity generated makes it crackle.
- When you walk across carpet, you generate static electricity. If you touch a metal object, like a door knob, the positive charge you have generated will leap to the uncharged doorknob—creating a spark and giving you a slight shock.
- The earliest known comb is believed to be the dried backbone of a large fish.
- Man-made combs dating back to 4000 B.C.E. have been found in Egyptian tombs.
- The British did not comb their hair until the Danish invaded in 789 C.E. and began teaching the English living on the coast to comb their hair regularly.

- In the 1600s, many Europeans erroneously believed that frequently combing grey hair with a lead comb could restore the hair to its original color.
- Static electricity is an electrical charge that cannot move through an object, but instead remains static. The object holds the *static* electrical charge—until it is touched by another object that conducts electricity.

IT'S ALL GREEK TO ME

The word *electricity* stems from the Greek word *elektron*, which means "amber." Why? The ancient Greeks discovered that rubbing a piece of amber (fossilized tree resin) with a cloth causes the amber (now charged with static electricity) to attract feathers.

Disappearing Chalk

WHAT YOU NEED

- ☐ 1 cup white vinegar
- ☐ Clean, empty mayonnaise or pickle jar
- ☐ 1 stick white chalk

WHAT TO DO

Pour the vinegar in the jar, drop in the stick of chalk, and wait ten minutes.

WHAT HAPPENS

Bubbles rise from the stick of chalk, which soon breaks into small pieces and dissolves completely.

WHY IT WORKS

The acetic acid in the vinegar dissolves the calcium carbonate in the chalk, releasing carbon dioxide gas.

BIZARRE FACTS

- ■ Chalk is soft, fine-grained, white limestone—made mostly from small shells and calcite crystals—that did not change into hard rock.
- ■ The White Cliffs of Dover in England are made from chalk.
- ■ When acid rain falls on statues containing limestone, the statue slowly

deteriorates, like the detail on the Parthenon in Athens, Greece.

- The chalk deposits of western Kansas contain preserved skeletons of extinct sea serpents, flying reptiles, birds, and fishes.
- Placing a piece of chalk in a jewelry box prevents rust by absorbing the moisture.

CHALK IT ALL UP TO ARCHEOLOGY

Most chalk was formed during the Cretaceous Period of time (beginning 130 million years ago and lasting 65 million years), named from the Latin word *creta*, meaning "chalk."

Disappearing Ink

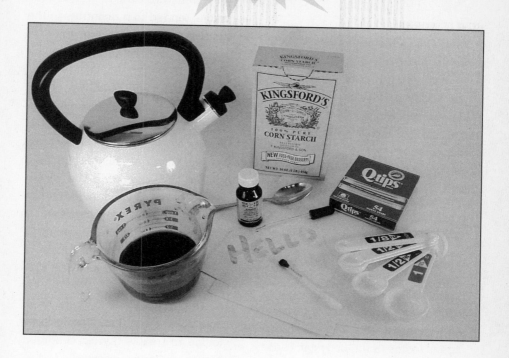

WHAT YOU NEED

- ☐ Teakettle
- ☐ Water
- ☐ Measuring cup
- ☐ Measuring spoons
- ☐ 1 teaspoon cornstarch
- ☐ Spoon
- ☐ Iodine
- ☐ Eyedropper
- ☐ Q-tip cotton swabs
- ☐ White paper

WHAT TO DO

With adult supervision, fill the teakettle with water and bring it to a boil. Carefully fill the measuring cup with one cup of boiling water. Add one teaspoon of cornstarch to the boiling water and stir with the spoon until completely dissolved. Remove any lumps. Using the eyedropper, add roughly ten drops of iodine to turn the mixture bluish, and stir well with the spoon. Use a Q-tip cotton swab to write your secret message or draw a picture on a piece of white paper. Let dry. (Note: Iodine is poisonous if swallowed. Do not drink or taste the iodine or the disappearing ink solution.)

WHAT HAPPENS

After several days, the blue ink disappears from the paper.

HOW IT WORKS

Adding the cornstarch to the water turns the liquid into a relatively strong base solution (pH 7.5). The iodine, when added to starch, turns blue. When the solution dries, carbon dioxide in the air mixes with the cornstarch solution, lowering the basicity enough to cause the iodine solution to vanish.

BIZARRE FACTS

- Kelp is rich with iodine.
- The thyroid gland in the human body produces iodine as part of a hormone called thyroxine, which controls the body's rate of physical and mental development. A shortage of iodine in the body can cause a goiter (an enlargement on the thyroid gland). To prevent iodine shortages in the body, manufacturers add iodine to salt, which they call iodized salt.
- Iodine was once commonly used as an antiseptic.
- Between 1944 and 1947, the nuclear weapons plant in Hanford, Washington, released radioactive iodine gas into the air. Cows grazed on grass contaminated by the airborne iodine, and when people drank the milk from the cows, the radioactive iodine tended to concentrate in the thyroid gland, in sufficient amounts to cause at least some cases of cancer.
- During the near meltdown of the Chernobyl nuclear power plant on April 26, 1986, a massive amount of radioactive iodine was released into the air. Cases of childhood thyroid cancer skyrocketed in the region because the thyroid gland tends to concentrate iodine ingested by the body. Poles were given iodine tablets to speed the elimination of radioactive iodine from their systems.
- In 1989, First Lady Barbara Bush was given a solution of radioactive iodine to treat a thyroid condition known as Graves' disease.

SEAWEED SALAD

Iodine was discovered in 1811 by French chemist Bernard Courtois, who located the chemical element in seaweed.

Dixie Cup Bridge

WHAT YOU NEED

- ☐ 75 5-ounce Dixie cups
- ☐ 3 sheets of corrugated cardboard (roughly 16-by-16 inches wide)

WHAT TO DO

Place twenty-five paper cups upside down on the floor in five rows of five cups each. Place the first sheet of cardboard on top of the cups.

Place another twenty-five paper cups upside down on top of the cardboard in five rows of five cups each. Place the second sheet of cardboard on top of the cups.

Place the last twenty-five paper cups upside down on top of the cardboard in five rows of five cups each. Place the third sheet of cardboard on top of the cups.

Slowly step on top of the cardboard.

WHAT HAPPENS

The paper cups support your weight.

WHY IT WORKS

Each paper cup is a cylinder, capable of supporting up to sixteen pounds of weight. The sheet of cardboard helps spread the weight equally across all the cups underneath.

BIZARRE FACTS

- Inventor Hugh Moore's paper cup factory was located next door to the Dixie Doll Company in the same downtown loft building. The word *Dixie* printed on the company's door reminded Moore of the story he had heard as a boy about "dixies," the ten-dollar bank notes printed with the French word *dix* in big letters across the face of the bill by a New Orleans bank renowned for its strong currency in the early 1800s. The "dixies," Moore decided, had the qualities he wanted people to associate with his paper cups, and with permission from his neighbor, he used the name for his cups.
- In 1923, Dixie cups produced a 2½ ounce Dixie cup for ice cream, giving the ice cream industry a way to sell individual servings of ice cream and compete with bottled soft drinks and candy bars.
- The Dixie Cups, a popular singing trio comprised of sisters Nadine, Marta, and Lucile LeCupsa, sang the 1964 hit song "Chapel of Love."
- While playing telephone operator Ernestine on *Saturday Night Live*, Lily Tomlin said, "Next time you complain about your phone service, why don't you try using two Dixie cups with a string?"

MY CUP RUNNETH OVER

In 1908, Hugh Moore started the American Water Supply Company of New England to market a vending machine that for one penny would dispense a cool drink of water in an individual, clean, disposable paper cup. Moore soon realized that his sanitary cups had greater sales potential than his water, particularly when Dr. Samuel Crumbine, a health official in Dodge City, Kansas, began crusading for a law to ban the public tin dipper. Lacking the capital to manufacture enough paper cups to abolish the tin dipper, Moore and his associate Lawrence Luellen traveled to New York City with a few handmade samples and eventually hooked up with an investment banker who invested $200,000 in the venture, incorporated as the Public Cup Vendor Company in 1909. That same year Kansas passed the first state law abolishing the public dipper, and Professor Alvin Davison of Lafayette College published a study reporting the germs of communicable diseases found on public dipping tins. As state after state outlawed public drinking tins, Moore and his associates created a paper cup dispenser to be distributed for free to businesses and schools who would buy the paper cups.

Duct Tape Lightning

WHAT YOU NEED

☐ Roll of duct tape

☐ Dark room

WHAT TO DO

On a cold day, peel three inches of duct tape from the roll and fold back an inch of tape so you are ready to pull a large piece of tape from the roll. Turn off the lights so you are in a pitch black room. Wait one minute until your eyes become accustomed to the dark. With your eyes peeled on the roll of tape, pull a long strip of duct tape from the roll quickly.

WHAT HAPPENS

You'll see a quick flash of light where the tape pulls free from the roll.

WHY IT WORKS

Stripping off the tape breaks the adhesive bond, causing the adhesive side of the tape to take on a negative charge, and the non-adhesive side to take on a positive charge. Electrons from the positive side leap through the air to neutralize the negative side, and as they collide with air molecules, they spark.

BIZARRE FACTS

■ Tim and Jim, the self-proclaimed Duct Tape Guys, have written four books about duct tape: *The Duct Tape Book, Duct Tape Book Two— Real Stories, The Ultimate Duct Tape Book* (subtitled "The third and final book in the Duct Tape Trilogy"), and *The Jumbo Duct Tape Book*. The Duct Tape Guys have also produced several *365 Days of Duct Tape* Page-A-Day calendars. You can visit the Duct Tape Guys on the Internet at www.ducttape. com.

- The astronauts aboard *Apollo 13* used duct tape to modify a carbon dioxide filter from the command module to fit into the lunar module, which the astronauts used as a lifeboat to return to earth.
- Canoe enthusiasts repair holes in their boats by applying a strip of duct tape on both the inside and the outside of the hole.
- In Scandinavia, some people call duct tape "Jesus Tape."
- Many contestants in the Miss America pageant use duct tape to enhance their figures during the evening gown and swimsuit competitions.

THE STICKY TRUTH

Cloth tape was originally invented for the American armed forces during World War II by the Permacel division of the Johnson & Johnson company. The military needed a durable, waterproof tape that could be ripped by hand and used to repair jeeps, planes, and to keep the moisture out of ammunition cases. Researchers at Johnson & Johnson gave cloth mesh a rubberized coating and applied a rubber-based adhesive to one side. During the housing boom in the United States that followed the war, contractor's used Johnson & Johnson's cloth tape to seal air-conditioning duct work and the gray tape became known generically as "duct tape."

Electric Waterfall

WHAT YOU NEED

- ☐ Clean, empty, plastic mayonnaise jar with a plastic lid
- ☐ Electric drill with a ¼-inch bit
- ☐ Black construction paper
- ☐ Scissors
- ☐ Black electrical tape
- ☐ Water
- ☐ Two hardcover books
- ☐ Kitchen sink
- ☐ Flashlight

WHAT TO DO

With adult supervision, drill a hole in the top of the plastic lid, one-half inch away from the edge. Drill a second hole on the opposite side of the lid, directly across from the first hole. Cut a sheet of black construction paper so you can roll it around the side of the jar and tape it securely in place, leaving the bottom of the jar exposed. Use the black electrical tape to cover up any remaining area of the jar where light might escape. Fill the jar with water and screw on the prepared lid tightly.

Turn out the lights in the room. Using the lit flashlight to see what you're doing, lie the jar down between two books, positioning the jar so that the holes in the lid are lined up vertically and a stream of water flows out

the bottom hole and into the sink. Hold the lit end of the flashlight against the bottom of the jar.

WHAT HAPPENS

The beam of light from the flashlight will bend as it travels through the curving stream of water flowing from the hole.

WHY IT WORKS

Although rays of light travel in a straight line, rays of light can be bent (or refracted). When light enters water at an angle, it bends because the speed of light is slower in water than in air. Some of the light enters the water, and some of the light is reflected back. When the curve of the water stream is less than the critical angle of 49° (as in this case), light is reflected back and forth between the surfaces of the water stream until it reaches the end and touches the sink. (The critical angle is different for different substances.) The water stream glows because some light emerges at ripples along the way.

BIZARRE FACTS

■ Light travels at the speed of 186,282 miles per second in a vacuum.

■ Light travels from the sun to the earth in about eight minutes.

■ A pencil in a glass of water appears to be broken at the waterline surface because of light refraction.

■ The amount that a ray of light bends when passing from one medium into another is called the *index of refraction*. Finding the index of refraction requires trigonometry. According to Snell's Law, developed by Dutch mathematician Willebrord Snell van Royen, the index of refraction equals the sine of the angle of incidence divided by the sine of the angle of refraction.

■ The average person consumes approximately 16,000 gallons of water in his or her lifetime.

■ The average American uses seventy gallons of water per day.

■ Water covers more than seventy percent of the earth's surface.

■ Wearing a mask underwater provides an air space so the swimmer's eyes can focus. However, when light changes speed going from water to the air inside the mask, the resulting refraction magnifies everything twenty-five percent, making things appear larger or closer.

Exploding Coca-Cola

WHAT YOU NEED

- ☐ 1 plastic cap from the top of a two-liter soda bottle
- ☐ Drill with ¼-inch bit
- ☐ 1 pack Mentos
- ☐ Nail (1 ¾-inch 5d finish)
- ☐ Dental floss
- ☐ Masking tape
- ☐ 1 full, two-liter bottle Coca-Cola

WHAT TO DO

With parental supervision, drill a ¼-inch hole through the center of the plastic bottle cap. Remove the plastic insert from inside the bottle cap.

Open the pack of Mentos and take out six candies. Gently punch a hole through the center of each of the six

Mentos candies by slowly pushing the nail through it, without cracking the hard candy coating. (Use the flat side of the plastic dental floss case to push the nail through the candy.)

Thread one end of the dental floss through the first Mentos candy and tie the dental floss to itself with two knots. Thread the free end of the dental floss through the remaining five

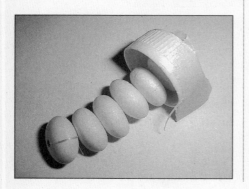

Mentos candies so they hang together like beads.

Thread the free end of the dental floss through the hole drilled in the bottle cap so the Mentos candies hang from inside the cap. Pull the free end of the dental floss taut and drape it over the outer end of the bottle cap and tape in place with a one-inch-long piece of masking tape (with one end of the tape folded back against its sticky side to create a tab).

Set the bottle of Coca-Cola on a flat surface outside and open the cap. Gently insert the strung Mentos into the neck of the bottle and screw the prepared cap securely in place without letting the Mentos touch the surface of the soda. Peel off the tape, allowing the string of Mentos candies to drop into the Coca-Cola, and quickly step back five feet.

WHAT HAPPENS
The Coca-Cola shoots from the hole in the bottle cap, creating a geyser seven feet high for approximately ten seconds, emptying half the soda from the bottle.

WHY IT WORKS
The gelatin and gum arabic from the dissolving Mentos candies weaken the surface tension of the water in the soda, allowing the carbon dioxide bubbles to expand. At the same time, the rough surface of the Mentos candies

lets new bubbles form more quickly (a process called nucleation). As more of the candies dissolve, both processes accelerate, rapidly producing foam. The resulting pressure inside the bottle forces a geyser of foam to spray from the bottle.

BIZARRE FACTS

- When cooking spaghetti in a pot of boiling water, organic materials leach out from the cooking pasta and weaken the surface tension of the water in the pot. This makes it easier for bubbles and foam to form, often causing the water to boil over. Adding a drop of vegetable oil to a pot of boiling water before adding the spaghetti strengthens the surface tension of the water and prevents it from boiling over.

- If you add a scoop of ice cream into a glass of root beer, the gums and proteins from the melting ice cream weaken the surface tension of the root beer, and the root beer foams over.

- If you drop a Mentos candy into a glass of flat soda, nothing will happen.

- Bookkeeper Frank M. Robinson, one of Coca-Cola inventor Dr. John Styth Pemberton's four partners, suggested naming the elixir after two of the main ingredients: the coca leaf and the kola nut. He suggested spelling kola with a *c* for the sake of alliteration. Robinson wrote the name in his bookkeeper's Spencerian script, much the way it appears today.

- Coca-Cola stock went public in 1919 at $40 per share. In 1994, one of those shares was worth $118,192.76, including dividends.

- If all the Coca-Cola ever produced were in regular-size bottles and laid end-to-end, they would reach to the moon and back 1,045 times. That is one trip per day for two years, ten months, and eleven days.

THE REAL THING

Since 1893, the recipe for Coca-Cola has been changed only once. In 1985, when Pepsi-Cola outsold Coca-Cola in the United States for the first time in history, the Coca-Cola Company sweetened the product and renamed it New Coke. Within three months, consumers forced the company to bring back the old formula. It became known as Coca-Cola Classic, and New Coke, considered the marketing fiasco of the decade, soon disappeared from the marketplace.

Exploding Egg

WHAT YOU NEED

- ☐ Clean, empty mayonnaise jar with lid
- ☐ 2 cups white vinegar
- ☐ Raw egg
- ☐ Turkey baster
- ☐ Water

WHAT TO DO

Fill the mayonnaise jar with vinegar, drop in the egg, and seal the lid. Let sit for three days or until the egg shell completely disintegrates, leaving only the membrane.

Using the turkey baster, drain all the vinegar from the jar, being careful not to poke the delicate membrane of the egg. Slowly fill the jar with water, reseal the lid, and let sit for several days.

WHAT HAPPENS

The egg will enlarge until the membrane bursts.

HOW IT WORKS

The acetic acid in the vinegar dissolves the calcium in the eggshell, making the shell disintegrate. The water, being a less concentrated solution than the egg's contents, passes through the semipermeable egg membrane (by osmosis), causing the pressure inside the egg to increase until the shell bursts.

BIZARRE FACTS

- If you squeeze the tapered ends of an egg between your palms, the egg will not crack open.
- Frogs do not drink water. They absorb water into their bodies by osmosis.
- Of all the birds in the world, the flightless kiwi of New Zealand lays the largest egg with respect to its body size. A kiwi egg may be up to five inches long and weigh more than a pound—twenty percent of the mother's weight.
- Hens lay up to 350 eggs or more in a year.
- The oyster produces up to five hundred million eggs a year.
- Emus lay eggs colored emerald green.
- Eggplants are named after the fact that the vegetable is shaped like a purple egg. At one time, the eggplant was thought to be poisonous.

EGGSISTENTIALISM

On the television series *Batman*, Vincent Price played Egghead, the smartest villain in the world (with an egg-shaped head), whose vocabulary included such exclamatories as "eggstraodinary," "eggceptional," and "eggciting."

Floating Egg

WHAT YOU NEED

- ☐ Clean, empty mayonnaise jar
- ☐ Hot tap water
- ☐ Canister of salt
- ☐ Tablespoon
- ☐ Raw egg

WHAT TO DO

Fill the jar halfway with hot tap water. Stir in one tablespoon of salt at a time until no more salt will dissolve in the water. Gently drop the egg into the salt water. The egg floats in the water. Now slowly pour regular tap water over the egg, filling up the rest of the jar.

WHAT HAPPENS

The egg floats on top of the salt water, but remains under the layer of regular tap water.

WHY IT WORKS

Welcome to Archimedes' Principle—the explanation behind buoyancy. The ancient Greek mathematician Archimedes stated that an object placed in a fluid is buoyed upward with a force equal to the weight of the fluid it displaces. In other words, adding salt to water increases the density of the water. The salt water is denser than the egg, so the egg floats. The egg, however, is denser than regular water, so the egg sinks in it. If left undisturbed, the egg will remain floating in the middle of the jar for several days.

BIZARRE FACTS

- ■ If you let the egg sit in the jar unrefrigerated for several days, the egg will go bad and float to the surface. (A bad egg, when broken, reeks; dispose of the bad egg carefully.)

- Greek mathematician Archimedes, who discovered the laws of the lever and pulley, also invented the catapult.
- White eggs and brown eggs are equally nutritious.
- Hard-boiling an ostrich egg requires forty minutes.
- In ancient Egypt, the apricot was called the "egg of the sun."

BAD EGG

If placed in a jar of regular tap water, a fresh egg will sink to the bottom because the egg is denser than water. If the egg has gone bad, however, it will float in the water. The bad egg floats because the yolk and albumen have dried up, making the egg less dense than water.

Flour Fireworks

WHAT YOU NEED

- ☐ 12-inch square of cheesecloth (available at grocery store)
- ☐ Measuring cup
- ☐ ½ cup white flour
- ☐ 12-inch piece of string
- ☐ Matches
- ☐ Candle
- ☐ Saucer

WHAT TO DO

Lay the piece of cheesecloth on a tabletop. Place the flour in the center of the cheesecloth square. Raise the

corners of the cheesecloth to make a bag and secure in place with the string.

With adult supervision, melt the bottom of the candle and secure it upright in the center of the saucer. Place the saucer on a cement surface outdoors. Light the candle. Gently shake the cheesecloth bag a foot above the candle flame, sprinkling a cloud of flour.

WHAT HAPPENS
The cloud of flour burns like sparkling fireworks.

WHY IT WORKS
Flour does not burn unless there is sufficient amount of oxygen between each particle. Shaking the cheesecloth filled with flour creates a cloud of flour dust oxidized sufficiently to ignite readily.

BIZARRE FACTS
- Since 1972, most of the flour found in home kitchens has been used for baking cookies.
- Mixing flour with water to a thick consistency creates glue.
- The average American eats approximately 125 pounds of flour every year.
- Ninety percent of the flour in America is made from wheat.
- People began making flour sometime between 15,000 and 9,000 B.C.E., using rocks to crush wild grain.

FLOUR POWER
During the European Renaissance, when the medieval ideal of feminine beauty required a woman's skin to be as white as a lily and her cheeks as red as a rose, peasants, unable to afford expensive cosmetics, made do with wheat flour and beet juice.

Flying Potato

WHAT YOU NEED

- ☐ Small red rose potato
- ☐ Two forks
- ☐ Toothpick
- ☐ Drinking glass

WHAT TO DO

Insert the tines of one fork into one end of the potato. Insert the tines of the second fork into the potato at the other end of the potato so that the handles of the forks are at a 45° angle from each other.

Insert the toothpick into the potato between the two forks.

Place the end of the toothpick on the edge of the drinking glass.

WHAT HAPPENS

The toothpick miraculously supports the weight of the forks and potato, keeping them balanced on the edge of the drinking glass. (If the potato does not balance on the toothpick, reposition the forks so the angle between the handles is smaller.)

WHY IT WORKS

The angle of the forks spreads their weight so that the center of gravity of the potato, forks and toothpick is concentrated on the toothpick.

BIZARRE FACTS

■ The word *fork* comes from the Latin word *furca*, meaning "pitchfork."

- Forks, usually made with only two tines, originated in Tuscany, Italy, in the eleventh century, but the clergy frowned upon their use, insisting that only human fingers should be used to eat God's bounty—along with spoons and knives.
- Forks became popular as eating utensils among French nobility during the eighteenth century.
- In the 1970s, China, having bought ten Boeing 707s and forty Pratt & Whitney replacement engines, ordered its aeronautical engineers to secretly build a copy of the Boeing 707 to be powered by one of the replacement engines. The plane, called the Y-10, could not fly. Chinese engineers accidentally mislocated the plane's center of gravity.
- A statue of Sir Francis Drake in Offenbach, Germany, wrongly proclaims the English explorer "Introducer of the Potato into Europe." There is no evidence that potatoes were aboard his ship, the *Pelican*.

FLY THE VERY FRIENDLY SKIES

To advertise its new leather first-class seats in Mexico, American Airlines literally translated its "Fly in Leather" slogan to *"Veula en Cuero,"* which means "Fly Naked" in Spanish.

Foam Factory

WHAT YOU NEED

- ☐ Measuring cup
- ☐ 1 can Barbasol shaving cream
- ☐ 4-ounce bottle of Elmer's Glue-All
- ☐ Large glass bowl
- ☐ Large spoon

WHAT TO DO

Fill the measuring cup with one cup Barbasol shaving cream. Spoon the shaving cream into the mixing bowl. Empty the bottle of Elmer's Glue-All into the mixing bowl. With the spoon, whip the glue and shaving cream together. Mold whatever you want out of the mixture. Let set overnight to dry.

WHAT HAPPENS

The mixture solidifies into foam.

WHY IT WORKS

Whipping the shaving cream and the glue together essentially fills the polyvinyl acetate molecules in the glue with air-filled soap lather. The glue dries filled with millions of tiny cells of air, much like foam rubber.

BIZARRE FACTS

■ Barbasol is a combination of the Roman word *barba* (meaning beard, and the origin of the word *barber*) and the English word *solution*,

CLOSE SHAVE

In 1931, Charles Goetz, a senior chemistry major at the University of Illinois, worked part-time in the Diary Bacteriology Department, improving milk sterilization techniques. Convinced that storing milk under high gas pressure might inhibit bacterial growth, Goetz began experimenting—only to discover that milk released from a pressurized vessel foamed. Realizing that cream would become whipped cream, Goetz began seeking a gas that would not saturate the cream with its own bad flavor. At the suggestion of a local dentist, Goetz succeeded in infusing cream with tasteless, odorless, nonflammable nitrous oxide, giving birth to aerosol whipped cream and aerosol shaving cream.

denoting the shaving cream is the same solution used by barbers. The stripes on the can evoke the familiarity of barbershop-pole stripes.

■ In 1920, Frank B. Shields, a former MIT chemistry instructor, developed the formula for Barbasol, one of the first brushless shaving creams on the market. Shields developed Barbasol especially for men with tough beards and tender skin because he had both of those shaving problems. The white cream in a tube—providing a quick, smooth shave—immediately won the allegiance of thousands, eliminating the drudgery of having to lather up shaving soap in a mug with a shaving brush and then rubbing it onto the face.

■ The original Barbasol factory and offices were both located in a small second-floor room in downtown Indianapolis. The tubes were filled with shaving cream, clipped, and packaged by hand. At the most, only thirty or forty gross made up an entire day's production schedule.

■ A 1937 advertisement for Barbasol read, "Barbasol does to your face what it takes to make the ladies want to touch it."

■ Shaving in the shower wastes an average of 10 to 35 gallons of water. To conserve water, fill the sink basin with an inch of water and vigorously rinse your razor often in the water after every second or third stroke.

■ According to archaeologists, men shaved their faces as far back as the Stone Age—20,000 years ago. Prehistoric men shaved with clamshells, shark teeth, sharpened pieces of flint, and knives.

■ Ancient Egyptians shaved their faces and heads during hand-to-hand combat so the enemy had less to grab. Archaeologists have discovered gold and copper razors in Egyptian tombs dating back to the fourth century B.C.E.

■ The longest beard, according to *Guinness Book of Records*, measured 17.5 feet long and was presented to the Smithsonian Institute in 1967.

■ Aerosol cans to deliver shaving cream were introduced in the mid-1950s.

■ The first shaving creams specifically targeted to women were introduced in 1986.

■ Seventy percent of women rate clean-shaven men as sexy.

■ Pfizer, which manufactures Barbasol, also makes Viagra.

■ The product known today as Elmer's Glue-All was first introduced by Borden in 1947 under the brand name Cascorez, packaged in two-ounce glass jars with wooden applicators. Sales did not take off until 1951 when Elsie the Cow's husband, Elmer, was chosen as the marketing symbol. In 1952, Borden repackaged Glue-All into the familiar plastic squeeze bottle with the orange applicator top.

■ Borden's Elmer's Glue operation in Bainbridge, New York, produces approximately 35 million four-ounce bottles of Elmer's Glue-All, School Glue, and GluColors annually.

Green Pennies

WHAT YOU NEED

- ☐ Three coffee filters
- ☐ Bowl
- ☐ White vinegar
- ☐ Ten pennies

WHAT TO DO

Place the three coffee filters, stacked together, in the bowl. Wet the coffee filters with the vinegar. Place the pennies on the coffee filter. Let sit for 24 hours.

WHAT HAPPENS

The pennies turn green.

WHY IT WORKS

The acetic acid in the vinegar combines with the copper coating on the pennies to form copper acetate, which appears green in color.

BIZARRE FACTS

- ■ The Statue of Liberty appears green because the acid in rain turns the copper statue into copper acetate.
- ■ A penny weighs more than a hummingbird.
- ■ The bumblebee bat of Thailand, the world's smallest mammal, weighs less than a penny.

- Nine pennies weigh exactly one ounce.
- "A bad penny" is someone undesirable, "a pretty penny" is a large sum of money, and "an honest penny" is money made honestly.
- Eighth-century English king Penda developed the coin and named it after himself.
- The female name Penny is a variation on the more formal name Penelope.
- A "penny pincher" is a cheapskate, a "penny wheep" is a small glass of beer, and a "penny ante" is a paltry sum of money.

A PENNY FOR YOUR THOUGHTS

Pennies are copper-coated zinc alloyed with less than three percent copper.

Homemade Chalk

WHAT YOU NEED

- ☐ Empty cardboard toilet paper tubes
- ☐ Scissors
- ☐ Waxed Paper
- ☐ Scotch Tape
- ☐ Plaster of Paris
- ☐ Water
- ☐ Bowl
- ☐ Spoon
- ☐ Tempera paints in various colors

WHAT TO DO

For each stick of chalk you wish to make, line the inside of an empty toilet paper tube with waxed paper and seal one end with tape.

With a spoon, mix two parts plaster of Paris with one part warm water in a bowl and add approximately two spoonfuls of tempera paint to achieve the desired color. Pour the mixture into the prepared toilet paper tubes. Gently tap the tube to release air bubbles from the plaster. Repeat for each color of chalk you wish to make.

Let the plaster mixture dry for forty-eight hours. Peel off the cardboard tube and wax paper.

WHAT HAPPENS

You've made colored chalk.

HOW IT WORKS

Chalk is made from limestone, the same ingredient in plaster of Paris.

BIZARRE FACTS

- Calcium carbonate, the main ingredient in chalk, is the key ingredient in toothpaste (like Colgate) and antacid tablets (like Tums).
- Plaster of Paris is made by heating crushed gypsum (hydrated calcium sulfate) until it has been dehydrated, leaving behind a white powder. Adding water to the powder causes a chemical reaction, changing the powder back to gypsum, which sets hard when the water evaporates.
- Plaster of Paris is naturally fire retardant. At roughly 600° Fahrenheit the water molecules stored in the plaster ooze out. This is why walls appear "sweaty" after a fire.

DON'T HAVE A COW, MAN

On the animated television comedy series *The Simpsons*, Bart Simpson was forced to write on a blackboard one hundred times at school, "I will not waste chalk."

Homemade Fingerpaints

WHAT YOU NEED

- ☐ Two mixing bowls
- ☐ Spoon
- ☐ One package Knox plain gelatin
- ☐ Measuring cup
- ☐ Water
- ☐ Pot
- ☐ ½ cup cornstarch
- ☐ Joy dishwashing liquid
- ☐ Six clean, empty, 4-ounce baby food jars
- ☐ Food coloring

WHAT TO DO

In a mixing bowl and using a spoon, mix the packet of powdered gelatin mix with ¼ cup water until dissolved. Set aside.

In a pot, mix the cornstarch with ¾ cup water. Add two cups hot water and mix well. With adult supervision, heat the pot on a stove, bringing the mixture to a boil while stirring constantly. When the mixture becomes clear and thick (after one to two minutes), remove the pot from the heat. Pour in the gelatin mixture. Mix well, then pour equal amounts of the mixture into the six baby food jars. Add one drop of Joy dishwashing liquid to each jar.

In the first jar, add five drops of yellow food coloring and stir well. In the second jar, add five drops of red food coloring and mix well. In the third jar, add five drops of green food coloring

and mix well. In the fourth jar, add five drops of blue food coloring and mix well. In the fifth jar, add four drops of yellow food coloring, one drop red food coloring, and mix well. In the sixth jar, add three drops red food coloring, two drops blue food coloring, and mix well. Let cool.

WHAT HAPPENS

You've created fingerpaints that can be used on heavy white paper. To store, seal the lids on the jars.

HOW IT WORKS

The cornstarch mixed with water is a hydrosol (a solid dispersed in liquid). Adding gelatin helps the cornstarch congeal and acts as an emulsion for the food coloring. The dishwashing

liquid breaks down the drops of fat in the gelatin, allowing the cornstarch and the gelatin to mix.

BIZARRE FACTS

- Fingerpaint can also be made by mixing food coloring with shaving cream, condensed milk, or plain yogurt.
- Chinese artists created fingerpaintings as early as 750 C.E.
- Fingerpainting is used as a form of therapy for the mentally ill, allowing them to express their feelings easily.
- As a food supplement, gelatin supplies the body with several amino acids lacking in wheat, barley, and oats.
- In 1845, Peter Cooper, inventor of the Tom Thumb locomotive, patented the first clear powdered gelatin mix. Fruit-flavored gelatin was invented in Le Roy, New York, fifty years later by carpenter Pearl B. Wait.
- Medicinal capsules are usually made from gelatin.

Hydrogen Balloon

WHAT YOU NEED

- ☐ Funnel
- ☐ 2 cups water
- ☐ Clean, empty wine bottle
- ☐ Rubber gloves
- ☐ Protective goggles
- ☐ 3 tablespoons Crystal Drano (sodium hydroxide)
- ☐ Aluminum foil (12-by-12 inches)
- ☐ Balloon
- ☐ String

WHAT TO DO

Working outdoors with adult supervision, use the funnel to pour the water into the wine bottle. Wearing rubber gloves and goggles, carefully add the Crystal Drano into the bottle. Carefully swirl the bottle to dissolve the Drano. Make twenty small balls from the aluminum foil (about ½ inch in diameter) and drop them into the bottle. Stretch the balloon well and immediately place the neck of the balloon over the mouth of the bottle. Let it inflate as big as it can get. This takes about ten minutes.

If the solution doesn't give off enough gas to fill the balloon, add more aluminum; if the glass gets too hot, you've used too much aluminum.

When the balloon is full, tie it off with the string. (Do not breathe the vapors from the bottle, and carefully dispose of the remaining solution in the bottle.)

WHAT HAPPENS

A chemical reaction gives off hydrogen gas, which fills the balloon. Since hydrogen is lighter than air, the balloon floats.

WHY IT WORKS

The aluminum interacts with the lye in the Crystal Drano, causing the hydrogen molecules to separate from the oxygen molecules in the water.

BIZARRE FACTS

- A box of Reynolds Wrap—the only nationally distributed brand of aluminum foil—can be found in three out of four American households.
- The *Hindenburg*, one of the largest airships ever built, burst into flames in 1937 over Lakehurst, New Jersey, when its hydrogen-filled bag exploded, killing thirty-six people. Today airships are filled with helium.
- The first hydrogen bomb, detonated in Enewetak Atoll in the Pacific Ocean in 1952, used fission to cause the fusion of the nuclei of two hydrogen atoms—yielding an explosion equivalent to ten million tons of TNT.
- Hydrogen is used as rocket fuel because the combustion reaction between hydrogen and oxygen propels the exhaust gas (primarily water vapor) out of the rocket's engine at 7,910 miles per hour-creating enormous thrust to lift the 4.4 million-pound space shuttle into orbit.
- If a hydrogen atom were the size of a golf ball, a golf ball would be the size of the earth.

UNSOLVED MYSTERY

On June 30, 1908, an explosion as powerful as a hydrogen bomb allegedly shook Siberia. Some witnesses reportedly saw a pillar of fire or a mushroom cloud. In 1927, Soviet scientists examined the twenty-five-foot-diameter site and determined that the charred earth had not been caused by a meteorite. Although many scientists today believe the crater was caused by an asteroid or comet, no one knows for certain.

Ice Cube Saw

WHAT YOU NEED

- ☐ Plastic Rubbermaid shoe box container
- ☐ Water
- ☐ Wire cutters
- ☐ 36-inch length of 22-gauge copper wire
- ☐ Two clean, empty bleach bottles with caps
- ☐ Tall, plastic garbage pail
- ☐ Plastic children's wading pool

WHAT TO DO

Fill the plastic shoe box container with water and place in a freezer for 24 hours to make a giant ice cube.

Using wire cutters, strip the coating of plastic insulation off the wire, leaving only copper wire. Tie one end of the wire to the handle of the first bleach bottle. Tie the other end of the wire to the handle of the second bleach bottle. Fill the two bleach bottles with water and seal tightly.

In a cool, shady spot, place the garbage pail upside down in the center of the plastic wading pool. Carefully remove the giant ice cube from the plastic shoe box container and place it on top of the upside down garbage pail.

Pick up the two bleach bottles (now wired together), guide the wire over the ice, and gently lower the bottles so they hang on the two sides of the garbage pail.

Let sit for several hours.

WHAT HAPPENS

The wire cuts through the ice cube, but the ice cube refreezes above it.

WHY IT WORKS

The pressure and the heat created by the friction of the wire from the weight of the bottles melts the ice. A small amount of water escapes the pressure by rising above the wire, but the cold temperature from the ice cube refreezes the water.

BIZARRE FACTS

- The average iceberg weighs twenty million tons.
- The white trail emitted from a plane is actually ice.
- Ice Cube's real name is O'Shea Jackson, and Vanilla Ice's real name is Robert Van Winkle.
- In 1851, surgeon John Gorrie of Apalachicola, Florida, built the first commercial ice-making machine.
- As water freezes into ice, the water molecules move apart and form a rigid pattern of crystals, expanding in volume by about one-eleventh.
- Ice floats because expansion makes it lighter than water.

THE ICEMAN COMETH

Approximately one-tenth of the earth's surface is permanently covered with ice. If all the ice melted, the sea would rise by about two hundred feet and many of the world's largest cities—including New York, Los Angeles, London, and Tokyo—would be underwater.

Liquid Crystal Magic

WHAT YOU NEED

- ☐ Measuring spoons
- ☐ 4.25-ounce packet hypo clearing crystals (available at photography store)
- ☐ Clean, empty glass mayonnaise jar with lid
- ☐ Teapot
- ☐ Measuring cup
- ☐ Spoon

WHAT TO DO

With adult supervision, measure ⅛ teaspoon of hypo clearing crystals and set aside for later.

Pour the packet of hypo clearing crystals into the jar. Bring the water to a boil in the teapot and using the oven mitt and measuring cup, pour 1½ cups of water into the jar. Screw the lid on the jar tightly and swirl the mixture around until all the hypo clearing crystals are dissolved (when there are no more grains on the bottom of the jar). The mixture will remain clear liquid. Let cool.

Drop the ⅛ teaspoon of crystals you set aside at the beginning of this project into the jar. Do not mix. Seal the lid on tightly and set the jar in a place where it will not be disturbed. Observe.

WHAT HAPPENS

Crystals begin to form in the glass, until all the liquid in the glass has changed into crystals.

WHY IT WORKS

Heating water allows it to dissolve more hypo than the water can normally hold. The cooled hypo solution is supersaturated. Adding just one crystal to the supersaturated solution disturbs the delicate balance of forces and starts the crystallization process. Once initiated, crystallization continues until the solution reaches a stable equilibrium.

BIZARRE FACTS

- Hypo crystals, the active ingredient in fixer that dissolves the unused silver salts out of developed photographs, are sodium thiosulfate.
- The quartz crystal in a quartz wristwatch vibrates 32,768 times per second.
- Comedian Billy Crystal's real last name is Crystal.
- Most nonliving substances—like metals and rocks—are made up of crystals.

LOOK INTO THE CRYSTAL BALL

The scientific study of crystals is called crystallography.

Magic Candle

WHAT YOU NEED

- ☐ Candle
- ☐ Matches
- ☐ Bowl
- ☐ Four pennies
- ☐ Clean, empty glass mayonnaise jar
- ☐ Blue food coloring
- ☐ 1 cup of water
- ☐ Spoon

WHAT TO DO

With adult supervision, melt the bottom of the candle and secure it upright in the center of the bowl.

Place four pennies equidistant from each other around the candle so the jar can sit over the candle with the jar rim resting on the pennies.

Add three drops of blue food coloring to the water and mix well with the spoon.

Pour the colored water into the bowl, light the candle, and set the jar over the candle so it sits on the pennies.

WHAT HAPPENS

The water rises roughly one-fifth of the way up the jar, the flame goes out, and the water remains in the jar.

WHY IT WORKS

Fire consumes the oxygen in the air. As the candle flame consumes the oxygen in the jar, the resulting pressure sucks the water into the jar to

replace the lost oxygen. When the lit candle consumes all the oxygen in the jar, the flame goes out.

BIZARRE FACTS

- Air is composed of twenty-one percent oxygen, seventy-eight percent nitrogen, and one percent trace elements.
- To human beings, air is invisible, odorless, and tasteless.
- People have lived more than a month without food and more than a week without water, but people cannot live more than a few minutes without air.

- Your right lung takes in more air than your left lung does.
- Air circulators in the Holland and Lincoln Tunnels under the Hudson River, connecting New Jersey and New York, circulate fresh air through the tunnels every ninety seconds.
- Before the invention of strings of electric Christmas lights, people decorated their Christmas trees with lit candles.
- Flamboyant pianist Liberace was known for his elaborate stage costumes studded with rhinestones and his grand piano decorated with an ornate candelabra.

BURNING THE CANDLE AT BOTH ENDS

In Sweden, on the morning of St. Lucia Day (celebrated on December 13) the oldest daughter in the home dresses in white, wears a wreath with seven lit candles on her head, and serves her family coffee and buns in bed.

Magnetic Ping-Pong Ball

WHAT YOU NEED

- ☐ Ping-Pong ball
- ☐ 1 foot of dental floss
- ☐ Scotch tape
- ☐ Kitchen sink

WHAT TO DO

Attach one end of dental floss to the Ping-Pong ball with a piece of Scotch tape. Turn on the water and, holding the free end of the string, let the ball hang under the stream of water running from the faucet.

WHAT HAPPENS

The ball sticks to the water, even if you hold the string on an angle.

WHY IT WORKS

According to Bernoulli's principle (named for Swiss mathematician Daniel Bernoulli), the water streaming rapidly over the top of the ball increases the air pressure under the ball, which then holds the ball up by air.

BIZARRE FACTS

- ■ Jim Henson made the original Kermit the Frog from a sleeve of his

ulate the styles of the best Ping-Pong players in the world and shoot a ball at up to sixty miles per hour.

- No American has ever won the men's world singles championship title in table tennis.

mother's winter coat, using two Ping-Pong balls for eyes.

- Cubans desperate to flee the communist country have used Ping-Pong paddles as oars to raft to Miami.
- Table tennis was originally played with paddles made from cigar box lids and balls made from champagne corks.
- America's table-tennis team trained for the 1992 Olympics in Barcelona with a $50,000 robot that could sim-

PING-PONG BOUNCES BACK

In the late 1880s, English engineer James Gibb, determined to come up with a game he could play indoors to get some exercise during the winter and rainy weekends, invented table tennis—a miniature version of tennis. Originally marketed under the name Gossima, table tennis did not take off until 1901, when a British manufacturer of table tennis equipment renamed the game Ping-Pong.

Milk Carton Engine

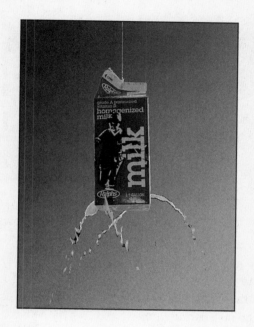

WHAT YOU NEED

- ☐ Sharpened pencil
- ☐ Clean, empty, 1-quart milk or juice carton
- ☐ Duct tape
- ☐ Nail
- ☐ String
- ☐ Water

WHAT TO DO

Using the pencil, punch a hole in the bottom left-hand corner of each of the four side panels of the milk or juice carton. Cut four square pieces of duct tape, fold back a flap on the end of each piece of tape to make the squares easy to remove, and adhere each square over one of the four holes in the carton. Using the nail, punch a hole in the top of the carton so you can tie one end of the piece of string through the hole. Use the other end of the string to hang the carton from a tree branch or a wooden beam outdoors. Fill the carton with water. Remove the four pieces of duct tape.

WHAT HAPPENS

As the water leaks from the four holes, the carton turns clockwise.

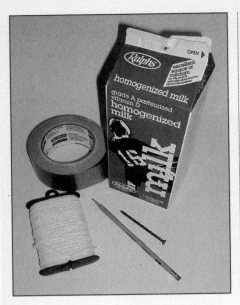

WHY IT WORKS

Sir Isaac Newton's third law of motion states, "For every action, there is an equal and opposite reaction." As the water pours from the holes, the reaction to it pushes the milk carton in the opposite direction. Since the carton is being pushed in four directions simultaneously, the carton spins.

BIZARRE FACTS

- Cows must eat grass to produce milk.
- A cow gives nearly 200,000 glasses of milk in its lifetime.
- In 1991, author P. J. O'Rourke told *Time* magazine, "This country is so urbanized we think low-fat milk comes from cows on aerobic exercise programs."
- The original Ford vehicles had Dodge engines.

THE GRAVITY OF THE SITUATION

In 1687, Isaac Newton published his theories of gravity, stating that the gravitational attraction between any two bodies in the universe is directly proportional to the two masses multiplied together and inversely proportional to the square of their distance apart. This theory, accepted by scientists for 240 years, was proved false by Albert Einstein.

Mirror Projector

WHAT YOU NEED

- ☐ Concave mirror (also known as a magnifying makeup mirror)
- ☐ Piece of white posterboard
- ☐ Room with a window

WHAT TO DO

Have an assistant stand one foot to the left of the window holding a sheet of white posterboard. Stand three feet from the window. Hold the mirror aimed to reflect the image of the window onto the posterboard.

WHAT HAPPENS

The image of the window is projected on the posterboard.

HOW IT WORKS

Light reflected from a mirror bounces back without spreading out. The rays of light from the window are reflected from the mirror, which acts like a lens to focus the rays of light, projecting a smaller image of the window on the posterboard.

BIZARRE FACTS

- ■ A mirror is usually made by adhering a thin layer of silver or aluminum onto a sheet of high-quality glass.
- ■ The ancient Egyptians, Hebrews, and Greeks developed the world's first mirrors by polishing metals like brass, bronze, silver, and gold.

- The superstition that a person who breaks a mirror receives seven years' bad luck originated in Rome during the first century C.E.
- In 1300, glass blowers in Venice, Italy, created the first glass mirrors.
- In *Through the Looking Glass*, Alice walks through a mirror.

MIRROR, MIRROR ON THE WALL

Rays of light bounce off the mirror at the same angle that they hit the mirror—the same way a pool ball bounces off the side of a pool table.

Monster Bubbles

WHAT YOU NEED

- [ ] Two 90-fluid-ounce bottles of Ultra Dawn dishwashing liquid
- [ ] Six 4-ounce bottles of glycerine (available at the drugstore)
- [ ] 1 gallon water
- [ ] Plastic bucket
- [ ] 12-inch square piece of cardboard
- [ ] Plastic children's wading pool
- [ ] Plastic holder from a six-pack of soda cans
- [ ] Two plastic drinking straws
- [ ] String
- [ ] Wire clothes hanger
- [ ] Hula hoop
- [ ] An eighteen-inch square plastic container
- [ ] Swimming mask

WHAT TO DO

Add the dishwashing liquid, water, and glycerine in the bucket. Swirl the ingredients gently to mix them without creating soapsuds. Cover the

67

bucket with the cardboard square and let the mixture sit undisturbed for five days.

Thread a three-foot length of string through two plastic drinking straws as if beading a necklace. Knot the ends of the string together and glide the knot inside one of the two straws.

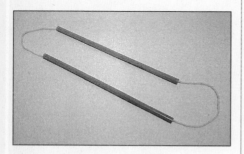

Bend the wire clothes hanger into a circle with a handle.

Tie four twelve-inch pieces of string to make four handles equidistantly around a hula hoop.

Place the plastic wading pool on a flat surface in the shade and away from any wind. Fill the pool with enough bubble solution to reach a depth of one inch.

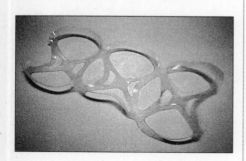

Dip the plastic holder from a six-pack of soda cans into the solution and use it to blow bubbles.

Hold the two plastic drinking straws apart so the string is taut, dip it in the bubble solution, then lift up while simultaneously bringing the straws together.

Submerge the wire clothes hanger into the bubble solution, then lift up, swishing through the air.

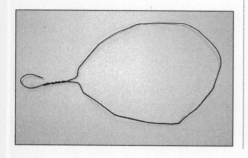

Place the hula hoop in the pool. Place the eighteen-inch square plastic container in the center of the pool. Have an assistant put on the swimming mask (to avoid getting soap in his or her eyes) and stand inside the

plastic container in the pool. With a second assistant, slowly lift the hula hoop from the pool and over the victim's head.

WHAT HAPPENS

You create dozens of extra-strength bubbles with the plastic holder from the six-pack of soda cans, you make large bubbles with the string of straws, you make enormous, monster bubbles with the wire coat hanger, and you create a bubble around a person with the hula hoop.

WHY IT WORKS

A soap bubble is a drop of water that has been stretched out into a sphere by using soap to loosen the magnetic attraction that exists between water molecules. Glycerine helps gives the walls of the bubble strength. When you wave the wire coat hanger through the air, for instance, the air pushes apart the molecules in the soapy film, but the molecules, attracted to each other, contract, forming the smallest surface possible to contain the largest volume of air possible—a sphere.

BIZARRE FACTS

- You can blow bubbles through a drinking straw or an empty toilet paper roll.
- The surface tension of a soap bubble is uniform for the entire bubble.
- Bubbles filled with carbon dioxide (blown from your mouth) last longer than bubbles filled with air.
- Bubbles made from a warm soapy solution last nearly twice as a long as bubbles made from a cold soap solution. Warmth sustains the surface tension of the bubbles. You can warm the bubble solution by placing it in a pot and heating it to 120° Fahrenheit.
- The more detergent used to make the bubble solution, the larger the

BUBBLE, BUBBLE, TOIL AND TROUBLE

Making monster bubbles successfully depends upon several variables, including air temperature and humidity. The more humidity in the air, the easier it is to make large bubbles. Bubbles also tend to burst quickly in direct sunlight.

bubbles will be. If you use more detergent than water (as instructed above), you can create monstrous bubbles.

■ Adding glycerine slows down the evaporation of the water in the bubble.

■ Bubbles blown on a rainy day last longer because of the moisture in the air.

■ By wetting one end of a plastic straw in bubble mix, you can gently push it through a large bubble and then blow a second bubble inside the first bubble.

■ The study of bubbles is called bubbleology.

■ The hula hoop gets it name from the Hawaiian hula because the gyrations made while rotating the hoop around one's waist match the movements made by doing the sensual hula dance.

■ The hula was originally a religious dance performed by Hawaiians to promote fertility.

■ You can make a Frisbee golf course by hanging a hula hoop from a tree branch. Designate a tee-off spot and toss the Frisbee toward the "hole," pick it up wherever it lands, and continue tossing until you get the Frisbee through the hula hoop. Set up nine different holes and keep score. The player with the fewest tosses to get through all the holes wins.

THE HOOPLA OVER THE HULA HOOP

The hula hoop craze swept across the United States in 1958, and within six months Americans purchased twenty million hula hoops. The hula hoop had actually been invented in ancient Egypt three thousand years earlier, where it was made from grapevines that had been dried and stripped.

Paper Cup Boiler

WHAT YOU NEED

- ☐ Candle
- ☐ Match
- ☐ Saucer
- ☐ Dixie cup
- ☐ Water
- ☐ Tongs

WHAT TO DO

With adult supervision, melt the bottom of the candle and secure it upright in the center of the saucer. Place the saucer on a cement surface outdoors. Light the candle.

Fill the cup halfway with water. Using tongs, hold the water-filled cup over the open flame, keeping the flame focused under the direct center of the cup (without making contact with the outer bottom rim).

WHAT HAPPENS

The water in the cup boils without the cup catching fire.

WHY IT WORKS

The temperature at which water boils (212° Fahrenheit) is significantly lower than the temperature at which paper burns (451° Fahrenheit), so the water cools the paper enough to prevent it from burning. (If the flame comes in contact with the bottom rim of the Dixie cup that does not make contact with the water, the rim and the cup may catch fire.)

BIZARRE FACTS

- The novel *Fahrenheit 451* by Ray Bradbury, titled for the temperature at which paper burns, is about a futuristic fireman who sets fire to banned books. The novel was made into a 1967 movie directed by François Truffat and starred Oskar Werner and Julie Christie.

- Davidson, Saskatchewan, is home to the world's largest coffeepot, measuring 24 feet tall and capable of holding 150,000 eight-ounce cups of coffee.

- The phrase "my cup runneth over" first appears in the Hebrew Bible in Psalm 23.

WET YOUR WHISTLE

The ceramic cups once used in British pubs had a whistle baked into the handle so patrons desiring a refill could whistle for service, giving birth to the phrase "Wet your whistle."

Potato Gun

WHAT YOU NEED

- ☐ PVC pipe (¾-inch in diameter and 12-inches long)
- ☐ Wooden dowel (⅝-inch in diameter and 12-inches long)
- ☐ Knife
- ☐ Cutting board
- ☐ Several large potatoes

WHAT TO DO

With adult supervision, use the knife and cutting board to cut the potatoes into slices one inch thick.

Press an open end of the PVC pipe down onto one of the potato slices like a cookie cutter to cut out a disk that remains in the tube like a cork. Repeat with the other end of the PVC pipe.

Aiming one end of the PVC pipe into the air, briskly push the wood dowel into the bottom end of the pipe.

WHAT HAPPENS

A potato cork shoots ten to fifteen feet from the pipe, making a funky popping sound.

WHY IT WORKS

When you push the wooden dowel into the PVC pipe, it pushes the first potato disk farther into the tube, compressing the air inside the tube. The increased air pressure forces the second potato disk to spring from the tube.

BIZARRE FACTS

- The potato originated in the Peruvian and Bolivian Andes mountains, where farmers cultivated it as early as 200 C.E.
- The Incas invented freeze-dried potatoes. They left potatoes out for several days, allowing them to continually freeze by night and thaw by day, then squeezing out the remaining moisture by hand, and then drying the potatoes in the sun.
- Spanish conquistadors first introduced the potato to Europe in the late sixteenth century.
- From 1847 to 1850, potato blight, a fungus disease, swept across Ireland, destroying the country's entire potato crop for four consecutive years. The resulting famine killed more than one million people. Since then, blight-resistant strains of the potato were imported from South America.

YOU SAY POTATO

In 1946, the first toy commercial aired on television. The toy advertised was Mr. Potato Head.

Psychedelic Milk

WHAT YOU NEED

- ☐ Paper plate
- ☐ Milk
- ☐ Food coloring (red, blue, green, and yellow)
- ☐ Dawn dishwashing liquid

WHAT TO DO

Pour the milk into the plate. Near the edge of the milk, add two drops of each color of food coloring. Add one drop of Dawn dishwashing liquid at the center of the milk.

WHAT HAPPENS

The food coloring begins to make wild psychedelic swirls, continuing to dance in the milk for about two minutes. Then the colors mix together and turn muddy gray.

HOW IT WORKS

At first the drops of food coloring in the milk remain separate because water-based food coloring does not mix with the fat in the milk, but at the same time, the surface water molecules in the milk pull on the puddles of color, spreading them equally in all directions. The dishwashing liquid weakens the pull of the water molecules in the center, causing the stronger water molecules along the rim of the plate to pull the puddles of color toward them. The dishwashing liquid breaks down the drops of fat in the milk, allowing the food coloring and the milk to mix.

BIZARRE FACTS

- Abraham Lincoln's mother died from drinking the milk of a cow that had eaten poisonous white snakeroot.
- The sap of the South American milk tree (*Brosimum utile*) looks, tastes, and can be used just like cow's milk.
- In 1982, Urbe Blanca, a cow in Cuba, produced 241 pounds of milk in one day—a world record. That's enough milk to provide 120 people with nearly a quart of milk.
- In 1937, Andy Faust of Collinsville, Oklahoma, milked 88.2 gallons of milk by hand in twelve hours—a world record.
- In the eighteenth and nineteenth centuries, unscrupulous food manufacturers used colorings to disguise spoiled foods.
- In 1856, Sir William Henry Perkins discovered the first synthetic dye, derived from coal-tar.
- The ancient Aztecs used cochineal, a red dye prepared from the dried bodies of female *Dactylopius coccus*, an insect which lives on cactus plants in Central and South America. Cochineal is still used today in food coloring, medicinal products, cosmetics, inks, and artists' pigments.
- In the United States, the first federal regulation concerning food colors was an 1886 act of Congress allowing butter to be colored.
- Studies show that people judge the quality of food by its color. In fact, the color of a food actually affects a person's perception of its taste, smell, and feel. Researchers have concluded that color even affects a person's ability to identify flavor.

COLOR MY WORLD

An extensive survey conducted by the National Academy of Sciences in 1977 estimated that the average American consumes 327.6 milligrams of FD&C color additives every day. That's sixteen times the Recommended Daily Allowance for iron. According to the survey, every day each American consumes an average of 100 milligrams of FD&C Red Dye Number 40, 43 milligrams of FD&C Yellow Dye Number 5, and 37 milligrams of FD&C Yellow Dye Number 6.

Racing Cans

WHAT YOU NEED

- ☐ Two coffee cans (13-ounce) with plastic covers
- ☐ Water
- ☐ Two pieces of ¾-inch pinewood, 1-by-3 feet
- ☐ Two books, both the same thickness (at least 1 inch)

WHAT TO DO

Fill the first coffee can halfway with water and seal the plastic lid. Fill the second coffee can to the top with water and seal the plastic lid.

Place one book on the floor. Prop up the end of one board on the books to create an incline. Construct a second incline next to the first.

Place each coffee can on its side at the top of the incline. Release the cans simultaneously and let them roll down the inclines.

WHAT HAPPENS

The filled can rolls faster than the half-filled can and comes to a stop. The half-filled can rolls to the same spot, then stops and rolls backward—rolling a greater distance than the full coffee can.

HOW IT WORKS

The rotational inertia of the half-filled can is greater than the rotational inertia of the full can. The full coffee can slows down due to the friction between the water and the inside of the can. The half-filled can rolls farther because the air inside allows the water to glide inside with less friction, giving the half-filled can greater rotational inertia and making it more difficult to stop.

BIZARRE FACTS

- English physicist Sir Isaac Newton first described inertia in 1687 in his first law of motion, which states that a body in motion remains in motion at a constant speed and in the same direction and a body at rest remains at rest unless acted upon by an outside force.

- Before Sir Isaac Newton died in 1727, his last words were, "I do not know what I may appear to the world; but to myself I seem to have been only like a boy playing on the seashore, and diverting myself in now and then finding a smoother pebble or prettier shell than ordinary, whilst the great ocean of truth lay all undiscovered before me."

Rain Machine

WHAT YOU NEED

- ☐ Two pots
- ☐ Twenty ice cubes
- ☐ 1 quart water
- ☐ Oven mitt

WHAT TO DO

Place the ice cubes in the first pot. With adult supervision, fill the second pot with water and bring it to a boil. Using the oven mitt, carefully hold the first pot six inches above the pot of boiling water. Observe the bottom of the first pot.

WHAT HAPPENS

Rain falls from the bottom of the pot filled with ice.

HOW IT WORKS

When steam rising from the pot of boiling water touches the cold bottom of the pot filled with ice, it changes back into water, forming droplets.

BIZARRE FACTS

■ In nature, water heated by the sun evaporates into water vapor, rises into the air, and cools, forming clouds, eventually becoming heavy enough to fall back to earth as rain.

- The place in the world with the greatest rainfall is Tutenendo, Colombia, with an average annual rainfall of 463.5 inches per year.

- No one knows who originated the saying "rain cats and dogs" or the nursery rhyme "Rain, rain, go away, come again another day."

- A typical raindrop falls at a speed of about seven miles per hour.

- The word *pluvial* means "pertaining to rain."

- Most Americans believe the saying "Neither snow nor rain nor heat nor gloom of night stays these couriers from the swift completion of their appointed rounds" is the motto of the U.S. Postal Service. However, the U.S. Postal Service does not have an official motto. The famous saying, written by the Greek historian Herodotus, is inscribed on the General Post Office in New York and describes Persia's mounted postal couriers of the fifth century B.C.E.

I DON'T MIND

The Beatles' song "Rain," released as a single in 1966, ends with lyrics that sound like gibberish. The strange mutterings are actually the vocal chorus run backward on a reel-to-reel tape deck—an effect added late at night by a drunken John Lennon in his private studio.

Rainbow Machine

WHAT YOU NEED

- ☐ Clean, empty 1-liter plastic soda bottle with screw-on cap
- ☐ Funnel
- ☐ Measuring cup
- ☐ 1 cup honey
- ☐ 1 cup Karo light corn syrup
- ☐ 1 cup water
- ☐ 1 cup rubbing alcohol
- ☐ 1 cup olive oil
- ☐ Food coloring

WHAT TO DO

Using the funnel, slowly pour the honey into the bottle. When the honey has settled to the bottom, tilt the bottle and slowly pour in the corn syrup, letting it dribble down the inside wall of the bottle to form a second layer on top of the honey. Add five drops of red food coloring to the water and stir well with the spoon. Tilt the bottle and slowly pour in the colored water, letting it dribble down the inside wall of the bottle to form a third layer. Follow the water with the olive oil. Add five drops of blue food coloring to the alcohol and stir well. Tilt the bottle and slowly pour the alcohol into the bottle. The layers float on top of each other without mixing. Secure the cap on the bottle and vigorously shake the bottle, mixing the contents together. Set the

bottle on a table and observe. Then let sit for eight hours.

WHAT HAPPENS

After being shaken, the layers mix together forming a purple liquid. Eight

81

hours later, the liquid separates into five distinct layers.

WHY IT WORKS

Each of the various liquids are immiscible, meaning they do not mix together. Shaking the bottle mixes the immiscible liquids together temporarily into an emulsion, but the emulsion soon separates again. The honey falls to the bottom because it is the densest liquid. The oil, being the least dense liquid, rises to the top layer. The food coloring remains mixed in the water because the food coloring is water soluble.

BIZARRE FACTS

- An emulsion is a mixture of one liquid evenly dispersed in the other—with tiny droplets of the dispersed liquid suspended in the other liquid. Photographic film is incorrectly said to be coated with an emulsion. The coating is actually a colloid.
- Adding an emulsifying agent to an emulsion prevents the immiscible liquids from separating. A few drops of dishwashing liquid added to a mixture of oil and water keeps the oil suspended in the water.
- Milk is an emulsion of butterfat in water. The emulsifying agent in milk is the protein casein.
- The proverb "More flies are taken with a drop of honey than a ton of vinegar" first appeared in *Gnomologia* by Thomas Fuller in 1732.
- The honeybee's distinctive buzz is actually the sound of its wings stroking 11,400 times per minute.
- Honey never spoils. It crystallizes, but if warmed in a microwave, returns to its liquid state.
- Corn syrup is made by cooking cornstarch and water under pressure and adding enzymes to the mixture—turning it into glucose, maltose, and dextrin. Another enzyme can be added to turn some of the glucose into fructose.
- Italy is the world's leading producer of olive oil, followed by Spain and Greece.
- Olive Oyl is the name of Popeye the Sailor Man's girlfriend.

HONEY DON'T

Utah is known as the beehive state, despite the fact that North Dakota leads the nation in honey production.

Self-Inflating Beach Ball

WHAT YOU NEED

- ☐ Electric drill with a ¼-inch bit and a ¾-inch bit
- ☐ Cork
- ☐ Eyedropper
- ☐ Uninflated beach ball (1 foot in diameter)
- ☐ Scissors
- ☐ Electrical tape
- ☐ Bucket
- ☐ Wooden mixing spoon
- ☐ 1 cup sugar
- ☐ 2 tablespoons molasses
- ☐ Packet of yeast (¼ ounce)
- ☐ 3 quarts water
- ☐ Funnel
- ☐ Clean, empty, 1-gallon Gatorade bottle

WHAT TO DO

With adult supervision, drill a ¼-inch hole through the middle of the cork. Insert the glass tube from the eyedropper through the hole in the cork so the tip protrudes from the top of the cork.

Insert the tip of the eyedropper into the nozzle on the uninflated beach ball and secure in place with a piece of electrical tape.

Drill a ¾-inch hole through the center of the lid of the Gatorade bottle. Insert the cork firmly in the hole in the lid.

In the bucket and using the wooden spoon, mix the sugar, molasses, yeast, and water until dissolved. Using the funnel, pour the mixture into the Gatorade bottle. Seal the prepared lid (complete with the cork and attached beach ball) onto the bottle. Set in a warm place for several days.

WHAT HAPPENS
The beach ball inflates with carbon dioxide.

HOW IT WORKS
The yeast is a plant that grows in the mixture, changing the sugar to carbon dioxide gas which fills the ball.

BIZARRE FACTS
- The resulting liquid in this experiment is a fermented mixture that can be made into rum through a complicated distilling process.
- The yeast used to ferment alcoholic beverages is actually a fungus. Members of the same fungus family

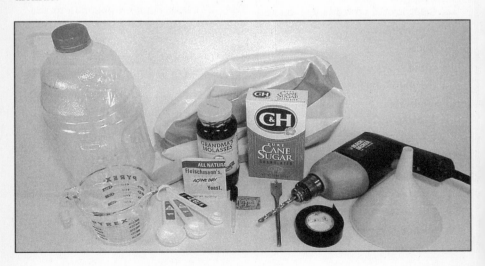

(Asomycetes) produce the antibiotics penicillin and streptomycin.

- In 1876, French scientist Louis Pasteur first reported that yeasts were living cells.
- Bakers add yeast to dough to make it rise. In bread-making, the yeast converts the sugar in the flour into alcohol and carbon dioxide. Bubbles of gas get trapped by the gluten in the dough, and as the gas expands, the bread rises. Baking destroys the yeast and causes the alcohol to evaporate from the bread.

Before yeast began being produced commercially in the 1880s, breadmakers prepared dough and left it uncovered so that airborne yeast plants landed on it and began the fermentation process.

MURDER BY BREATHING

If you are locked in a sealed room, you will die of carbon dioxide poisoning—before all the oxygen is depleted from the room.

Silver Egg

WHAT YOU NEED

- ☐ Candle
- ☐ Matches
- ☐ Saucer
- ☐ Tweezers
- ☐ Piece of eggshell
- ☐ Glass of water

WHAT TO DO

With adult supervision, melt the bottom of the candle and secure it upright in the center of the saucer. Light the candle.

Using the tweezers, hold the eggshell over the flame until it is smoked black. Then submerge the eggshell in the glass of water.

WHAT HAPPENS

The eggshell appears to be silver.

HOW IT WORKS

The flame deposits lampblack and small amounts of cracked paraffin on the eggshell. This coating is a hydrophobic mixture, meaning it repels water. Thousands of tiny air bubbles accompany the submerged eggshell, preventing the black coating from getting wet. Light reflects from these air molecules—making the eggshell appear silver.

BIZARRE FACTS

- ■ In psychology, the word *hydrophobic* means "fear of water," but in

chemistry the same word means "having no affinity for water."

■ No word in the English language rhymes with silver.

■ A small group of mammals known as monotremes lay eggs with a leathery shell. The only monotremes are the echidnas (five species of spiny anteaters) and the duck-billed platypus—all indigenous to Australia, New Guinea, and Tasmania.

■ Eggs Benedict was named after socialite Samuel Benedict, who, suffering from a hangover one morning in 1894, asked the maître d'hôtel in New York City's Waldorf-Astoria Hotel for bacon and poached eggs on toast with Hollandaise sauce. The maître d' substituted ham for the bacon and an English muffin for the toast, creating a new breakfast sensation. Benedict Arnold (1741–1801), the most famous traitor in American history, never ate eggs Benedict.

■ The United States government stores its supply of silver at the U.S. Military Academy at West Point, New York.

■ In Frank L. Baum's classic children's book *The Wonderful Wizard of Oz*, Dorothy wears silver shoes. Hollywood screenwriter Noel Langley changed them to ruby slippers in the script for MGM's classic 1939 movie, *The Wizard of Oz*.

■ In 1876, Nell Saunders, winner of the first United States women's boxing match, received a silver butter dish as a prize.

■ More silver is mined in Mexico than any other country, followed by Peru and Russia.

■ The sixteenth-century astronomer Tycho Brahe, having lost his nose in a duel with one of his students, wore an artificial nose made from silver.

DON'T CRY FOR ME, ARGENTINA

The chemical symbol for silver is Ag, derived from the Latin word for the metal, *argentum*, from which the country Argentina gets its name (because the first Spanish settlers came in search of silver and gold, which they failed to find).

Slinky Race

WHY IT WORKS

A Slinky sitting at the top of a staircase has *potential energy*. When a force is applied to the Slinky to make it start down the stairs, the Slinky is affected by gravity. The potential energy becomes *kinetic energy*, and the Slinky flips coil over coil down the stairs. As the Slinky flips down the steps, the energy is transferred along its length in a compressional wave (resembling a sound wave).

How fast a Slinky walks down steps depends how quickly a longitudinal wave travels through the Slinky coil, which depends on the tension and mass of the coil. The tighter the coil, the faster the longitudinal wave travels through the Slinky. The smaller the mass of the Slinky, the tighter the tension in the coil. The wave moves faster through the smaller Slinky, making it travel quicker.

WHAT YOU NEED

- ☐ **Traditional metal Slinky**
- ☐ **Metal Slinky Jr.**
- ☐ **Staircase**

WHAT TO DO

Place a Slinky and a Slinky Junior at the top of a staircase. Simultaneously flip the top coil of each Slinky to the next lower step and let go.

WHAT HAPPENS

The smaller Slinky beats the larger Slinky down the stairs.

BIZARRE FACTS

- ■ Slinkys adorn lighting fixtures in Harrah's Casino in Las Vegas because the interior designers like the unusual shadows the Slinkys cast.
- ■ United States soldiers using radios during combat in Vietnam tossed

the Slinky into trees to act as a makeshift antenna.

- The Slinky helps scientists understand the supercoiling of DNA molecules. Slinky and Shear Slinky, two computer graphics programs developed at the University of Maryland, use a Slinky model to approximate the double helix coiling of DNA molecules.

- In 1985, Space Shuttle astronaut Jeffrey Hoffman became the first person to play with a Slinky in zero-gravity physics experiments in orbit around the earth.
- Every Slinky ever made has been produced on the original machinery in the tiny town of Hollidaysburg, Pennsylvania.
- If you hold one end of a Slinky and whirl it around your head, it swings out from you. This is caused by centrifugal force. The faster you make the Slinky go around, the longer the centrifugal force stretches out the Slinky and raises it from the ground.
- Slinkys are used in physics classes to teach students the properties of waves.
- A Slinky can be seen in the movies *Demolition Man*, *Hairspray*, *The Inkwell*, *Other People's Money*, *The Pink Panther*, and *Ace Ventura 2: When Nature Calls*.
- In 1999, the United States Postal Service introduced the world's first Slinky stamp.
- Pecan harvesters have used Slinkys in machinery to help collect pecans.
- In 1943, Richard James, a twenty-nine-year-old marine engineer working in Philadelphia's Cramp Shipyard, tried to figure out how to use springs to mount delicate meters for testing horsepower on battleships. A torsion spring fell off his desk and tumbled end over end across the floor. Convinced he could devise a steel formula that would give the spring the right tension to "walk," James decided to make a toy out of his accidental discovery.
- Richard James' wife, Betty, named the spring toy Slinky after thumbing through the dictionary, because the word was defined as "stealthy, sleek, and sinuous."
- In 1960, Slinky inventor Richard James abandoned his business and family to join a religious cult in Bolivia, leaving his wife behind with six children to raise, a floundering business, and a huge debt. James died in Bolivia in 1974.

Smoking Fingertips

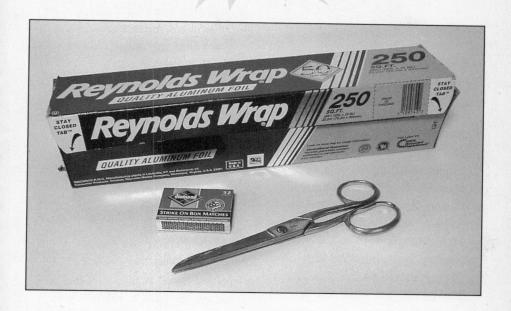

WHAT YOU NEED

- ☐ Scissors
- ☐ Box of safety matches
- ☐ Aluminum foil

WHAT TO DO

With scissors, cut the striking surface from one side of the box of matches. Place the striking surface facedown on a piece of aluminum foil, and with adult supervision, use a lit match to set the striking surface on fire, letting it burn until it turns to ashes, leaving an orange rust on the foil. Let cool.

Discard the ashes from the cardboard, and rub the tip of your index finger in the orange chemical on the foil, then rub your index finger and thumb together briskly.

WHAT HAPPENS

White smoke rises from your fingertips.

HOW IT WORKS

The striking surface on a box of safety matches contains red phosphorus, which is deposited on the aluminum foil after burning the strip. The warmth from the friction created by rubbing your fingers together causes the phosphorus to bond with the oxygen in the air to form phosphorous oxide—a white vapor.

BIZARRE FACTS

- The fingerprints of koala bears are almost indistinguishable from the fingerprints of human beings.
- During World War II, the United States Fleet Commander in the Pacific Theater (Admiral Chester Nimitz) and the Japanese Fleet Commander (Admiral Isoroku Yamamoto) had each lost two fingers as the result of accidents while they were younger officers aboard ships.
- Actor James Doohan, best remembered as Lt. Commander Montgomery Scott on the television series *Star Trek*, is missing the entire middle finger of his right hand.
- Grateful Dead guitarist Jerry Garcia was missing the top two-thirds of the middle finger on his right hand.

- The white, crescent shaped part of your fingernail is called the lunula (from the word *luna*, meaning "moon").
- Working in pressurized space suits causes the fingerprints on the fingertips of the Space Shuttle astronauts to be rubbed away.
- On the popular television comedy variety series *Rowan & Martin's Laugh-In*, the cast routinely bestowed The Flying Fickle Finger of Fate Award.

AT YOUR FINGERTIPS

The phrase "burn your fingers" means "suffering injury by meddling;" the phrase "have a finger in the pie" means "to have a share in something;" the phrase "keep your fingers crossed" means "to hope for good luck;" the phrase "put your finger on it" means "to recall or locate something;" the phrase "not lift a finger" means to "refrain from taking action;" the phrase "put the finger on someone" means "to designate a victim or identify a criminal;" and the phrase "slip through your fingers" means "to let an opportunity pass you by."

Snot on a String

WHAT YOU NEED

- ☐ Teakettle
- ☐ Water
- ☐ Oven mitt
- ☐ Clean, empty, glass mayonnaise jar
- ☐ Teaspoon
- ☐ Epsom salt
- ☐ Two drinking glasses
- ☐ 16-inch-long string of wool
- ☐ Two metal washers (approximately 1 inch in diameter)
- ☐ Saucer

WHAT TO DO

With adult supervision, fill the teakettle with water and bring to a boil.

Wearing an oven mitt, carefully fill the mayonnaise jar three-quarters full with boiling water. Still wearing the oven mitt, stir in one teaspoon of Epsom salt at a time until no more will dissolve. This takes approximately fifty teaspoons of Epsom salt. The resulting liquid is called a *saturated solution.*

Fill the two drinking glasses with the saturated solution. On a tabletop in a warm, secure place, set the two glasses on opposite sides of the saucer.

Tie each end of the wool string to its own metal washer. Drop each washer into its own glass of solution

so that the wool string hangs taut across the two glasses, like a high wire. Let sit for one week.

WHAT HAPPENS
A stalactite forms in the middle of the wool string, hanging over the saucer like frozen snot from a runny nose.

HOW IT WORKS
The string of wool absorbs the saturated Epsom salt solution, which slowly drips from the middle of the string to the saucer below. As the water evaporates, the atoms in the Epsom salt draw close together, forming a hanging column from the string—similar to the way stalactites hang down from the ceilings of caves or icicles hang from the eves of a roof.

BIZARRE FACTS
- Stalactites form when water drips through cracks in the roof of a cave and carries the calcium carbonate with it. As the water evaporates, a calcite column is left hanging from the ceiling of the cave. The formations that build up from the drips on the floor are called stalagmites. When stalactites and stalagmites join together, they form what look like stone curtains.
- The longest freehanging stalactite in the world is in the Poll, an Ionain cave in County Clare, Ireland. The stalactite measures twenty feet, four inches.
- The word *snot* also means "brat" and the adjective *snotty* means "arrogant."
- Epsom salt is actually magnesium sulfate in powder form. Epsom salt gets its name from the springs in Epsom, England, where the chemical was first mined.
- Epsom salt can be used to make crystal paintings. Simply draw with crayons on construction paper, then (with adult supervision) mix together equal parts Epsom salt and boiling water. Using a paintbrush, paint the picture with the salt mixture. When the salt solution dries, frosty crystals will appear.

Sonic Blaster

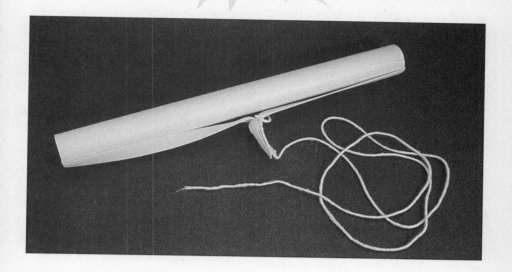

WHAT YOU NEED

- ☐ PVC pipe (3 ¼ inch in diameter and 12 inches long)
- ☐ 20-inch length of clothing elastic
- ☐ 5-foot length of string

WHAT TO DO

Thread one end of the elastic through the PVC tube and tie the two ends together securely to form a loop.

Tie one end of the string around the elastic at one of the tube openings.

Hold the free end of the string and whirl the PVC tube around your head, like a lasso. (You may wish to wear a bicycle helmet to avoid hitting yourself in the head with the PVC tube.)

WHAT HAPPENS

The PVC tube makes a loud roaring sound as it flies through the air.

HOW IT WORKS

As the tube spins around, air enters the tube, causing the elastic band to vibrate rapidly. This sound is carried out by the air leaving the tube.

BIZARRE FACTS

- The longer the length of string used, the louder the resulting sound.
- In the movie *Crocodile Dundee*, actor Paul Hogan makes a similar sonic blaster using a rock and a long piece of twine.

- The sound of knuckles cracking is caused by imploding synovial fluid, the liquid that keeps the joints moist and lubricated.
- When a bullwhip is cracked, the tip moves so fast that it actually breaks the sound barrier, creating a tiny sonic boom.
- Cats can hear ultrasound.
- Novelist William Faulkner titled his novel *The Sound and the Fury* after a line from Shakespeare's play *Macbeth*: "Life's but a walking shadow, a poor player that struts and frets his hour upon the stage, and then is heard no more; it is a tale told by an idiot, full of sound and fury, signifying nothing."
- One of the most popular songs recorded by the musical duo Paul Simon and Art Garfunkel is "The Sound of Silence."
- Any sound louder than 120 decibels is painful to the human ear. Racing cars emit 125 decibels. Rock concerts typically reach 130 decibels.
- Sound travels one mile through air in approximately five seconds. Sound travels one mile through water in approximately one second.

SOUNDS STRANGE

When the pilot of a plane breaks the sound barrier, the pilot does not hear the sonic boom heard by people on the ground (because the pilot is traveling faster than the speed of sound).

Spinning CD

WHAT YOU NEED

- ☐ 3-foot-long piece of dental floss
- ☐ Toothpick
- ☐ Clean, unneeded compact disc
- ☐ Scotch tape

WHAT TO DO

Tie one end of the piece of string to the middle of the toothpick. Insert the toothpick through the hole in the compact disc so that when you hold the other end of the string, the disc rests on the toothpick and the knot in the dental floss is in the center of the hole. Use two pieces of Scotch tape to secure the ends of the toothpick to the compact disc.

Try to swing the compact disc back and forth like a pendulum, keeping it level without letting it flop around.

Stop. Use your index finger to spin the compact disc on the string (as if tossing a Frisbee). Now try swinging it back and forth like a pendulum.

WHAT HAPPENS

As you swing the spinning compact disc back and forth, the disc remains level without any effort.

HOW IT WORKS

The spinning disc is a gyroscope, a simple machine that seems to defy the laws of gravity. A spinning body tends to spin on the same plane on which it began spinning, unless an outside force moves it from that original plane. This ability of a spinning body to always spin on the same plane is called *gyroscopic inertia*.

BIZARRE FACTS

- Gyroscopes guide airplanes and rockets, control stabilizers on ships, and can be used as a reliable navigational aid.

- The spinning wheels of a moving bicycle turn the vehicle into a gyroscope, allowing the rider to stay balanced with minimal effort.
- A spinning gyroscope is unaffected by the earth's gravity or by magnets.
- A spinning gyroscope holds its original position in space while the earth turns under it. If you point the axle of a spinning gyroscope at the sun, the axle appears to follow the sun as it crosses the sky.
- The first gyroscope in recorded history was built by German scientist G. C. Bohnenberger.
- In 1852, French physicist Jean Foucault built a gyroscope to demonstrate that the earth rotates on its axis. Foucault also named the device after the fact that he used it to view the revolution of the earth, combining the Greek word *gyros* (meaning "revolution") with the Greek word *skopein* (meaning "to view").

THE DIGITAL BOSS

The first audio CD manufactured in the United States was Bruce Springsteen's *Born in the USA*.

Swimming Pool Smoke Bomb

WHAT YOU NEED

- ☐ Clean, empty, 32-fluid-ounce Gatorade bottle
- ☐ 25 marbles
- ☐ Hot water
- ☐ Red food coloring
- ☐ Swimming pool

WHAT TO DO

Place the marbles inside the Gatorade bottle. Fill the Gatorade bottle with hot water. Add twenty drops of red food coloring, screw the cap on the bottle tightly, and shake well. Remove the cap, and drop the bottle into the deep end of a swimming pool.

WHAT HAPPENS

The bottle sinks to the bottom of the pool and the red liquid in the bottle steams up into the pool water, appearing as if a smoke bomb is going off in the bottom of the pool.

HOW IT WORKS

Heat rises. The warm red water in the bottle is lighter than the cold water of the pool because the molecules in hot water are farther apart than the molecules in cold water.

BIZARRE FACTS

- Rolling Stone Brian Jones drowned in a swimming pool.
- In the novel *The Great Gatsby* by F. Scott Fitzgerald, Jay Gatsby is shot while lying on an inflatable raft in his swimming pool.
- There are more swimming pools in Los Angeles than any other city in the world.
- The largest swimming pool in the world is in Casablanca, Morocco. The Orthleib Pool measures 1,574 feet long and 246 feet wide (an area of 8.9 acres) and is filled with seawater.
- All students at Cornell University must swim four laps across the pool in Teagle Gym before they are allowed to graduate.

THE HEAT IS ON

Until the late 1700s, many scientists mistakenly believed that heat was an invisible fluid called "caloric." They insisted that an object became warm when caloric flowed into it and grew cold when caloric flowed out of it. Since an object weighed the same whether it was hot or cold, scientists believed that caloric was weightless and could not be considered matter.

Underwater Blue Light Beam

WHAT YOU NEED

- ☐ Sharpened pencil
- ☐ Two cardboard squares (3-by-3 inches)
- ☐ Scissors
- ☐ Black electrical tape
- ☐ Empty cardboard toilet paper tube
- ☐ Flashlight
- ☐ Fishbowl with two flat sides
- ☐ Measuring cup
- ☐ Water
- ☐ Milk
- ☐ Eyedropper
- ☐ Wooden spoon
- ☐ Dark room
- ☐ Flashlight

WHAT TO DO

With the sharpened pencil, poke a hole in the center of both cardboard squares.

Using scissors, cut small strips of black electrical tape to adhere the two cardboard squares over the ends of the cardboard toilet paper tube. Then

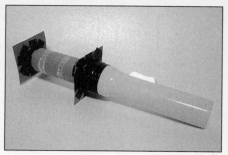

place one of the cardboard squares over the face of the flashlight and secure into place with strips of electrical tape. Wrap some electrical tape around the circumference of the head of the flashlight to prevent light from shining through the plastic or seeping through any cracks in the cardboard other than the punched hole.

Fill the fishbowl with water, add twenty drops of milk, and stir well with the wooden spoon.

In a dark room, hold the lit flashlight six inches away from one of the flat sides of the fishbowl so the beam of light shines into the bowl. Look into the top of the bowl.

WHAT HAPPENS

A misty blue beam of light can be seen passing through the milky water. The beam of light emerging from the other side of the fishbowl appears to be orange.

HOW IT WORKS

The beam of light is reflected and scattered in the water by the tiny particles of suspended fat from the milk, causing the beam of light to appear blue. This is called the Tyndall Effect, after British physicist John Tyndall. The light appears blue when the particles are smaller in diameter than one-twentieth the wavelength of light. The light emerging from the bowl appears orange because the blue light waves are trapped in the fishbowl.

BIZARRE FACTS

■ In 1976, British physicist John Tyndall observed that a *Penicillium* mold slowed the growth of bacteria—fifty years before Sir Alexander Fleming's chemical work on penicillin.

■ The Tyndall Effect is named after British scientist John Tyndall because he made the first comprehensive study of the effect, not because he discovered it. The effect was first witnessed in 1857 by English physicist Michael Faraday, best known for discovering the principle of electromagnetic induction.

■ You can also witness the Tyndall Effect when car headlights cut through the fog, when a flashlight beam is aimed through a room filled with dust or smoke, when sunlight comes in through a window and illuminates floats of dust, when beams of sunlight come through the clouds and illuminate the small drops of moisture in the air, or in a movie theater when the beam of light from the movie projector illuminates dust in the theater.

- The Tyndall Effect can be seen in most episodes of *The X-Files*, whenever Moulder and Sculley investigate a dark place with flashlights.
- The Tyndall Effect is used to determine whether a liquid is a suspension or a solution. The particles in a suspension are large enough to reflect or scatter light, making the light beam visible. A solution does not yield the Tyndall Effect.
- Several prominent scientists predicted that in January 1974 a comet named Kohoutek with a tail fifty million miles long would light up the sky, glowing brighter than the moon. The comet appeared as a faint streak of light in the sky near Venus and Jupiter.
- In the 1600s, scientists claimed that light travels through an invisible, weightless, frictionless, stationary, omnipresent substance called ether that filled all space. Centuries later, in 1905, Albert Einstein published his theory of relativity, showing that light does not rely on the existence of ether.

THE BLUE LIGHT SPECIAL

On December 1, 1903, the Iroquois Theater opened in Chicago, Illinois, billed as the world's first fireproof theater. That same month, a blue stage light blew out and set fire to the scenery, burning the "fireproof" theater to the ground.

Waterfall Weirdness

WHAT YOU NEED

- ☐ Sharpened pencil
- ☐ Clean, empty, 1-quart milk or juice cartoon
- ☐ Ruler
- ☐ 7-inch strip of masking tape
- ☐ Water
- ☐ Kitchen sink

WHAT TO DO

Using the sharp pencil, poke a hole in the bottom left-hand corner of one side of the milk carton, one inch from the bottom and one inch from the left-hand side. Poke a second hole three-and-a-half inches from the bottom and two inches from the left-hand side in the same side of the milk carton. Poke a third hole six inches from the bottom and three inches from the left-hand side in the same side of the milk carton.

Adhere the strip of masking tape over the three holes, folding over a small piece of one end of the tape to make a tab for easy removal.

Fill the milk carton with water. Stand the milk carton next to a sink,

with the side with the three holes toward the basin. Peel off the piece of masking tape. Observe.

WHAT HAPPENS

The water pours out in three streams of different projectile lengths. The top stream trickles out closest to the milk carton, while the bottom stream pours out furthest from the milk carton.

HOW IT WORKS

The weight of the water causes varying degrees of pressure in the milk carton. The deeper the water, the greater the pressure. The pressure causes the water at the bottom of the container to shoot out of the hole, while the water at the top of the container trickles out.

BIZARRE FACTS

- Water is approximately eight hundred times denser than air.
- Angel Falls, twice the height of the Empire State Building and twenty times as high as Niagara Falls, is named after Jimmy Angel, a U.S. pilot who discovered them in 1935. Angel landed his G-2-W Flamingo *El Rio Caroni* on top of Auyan Tepuy in 1937, where it remained for the next forty-three years, embedded in a bog.
- Grand Rapids, Michigan, was the first city in the United States to add fluoride to its water.
- In the 1995 movie *Waterworld*, the earth is completely submerged underwater, yet all the people, living in small boats on the water in the sun, are covered with dirt and grease.

TRUSTING THE MAN WITH THE STAR

In 1980, Texaco began drilling for oil from a new rig in the middle of Lake Peigneur in Louisiana. The water immediately drained from the 1,300-acre lake, sucking eight tugboats, nine barges, five houses, a mobile home, and two oil rigs into the abandoned salt mine beneath the lake.

Wax Snowflakes

WHAT YOU NEED

- ☐ Bowl
- ☐ Cold water
- ☐ Blue food coloring
- ☐ Spoon
- ☐ Candle
- ☐ Matches

WHAT TO DO

Fill the bowl with cold water, add five drops of blue food coloring, and stir well. With adult supervision, light the candle and hold it two feet above the bowl of water, allowing hot wax to drip into the water.

WHAT HAPPENS

The drops of wax float on the surface of the water and look like snowflakes.

HOW IT WORKS

The moment the liquid hot wax splatters on the surface of the cold water, the wax immediately solidifies, suggesting the appearance of snowflakes.

BIZARRE FACTS

- When the Coca-Cola Company introduced Coke in China, the name was transliterated as "Kek-

ter so the verse "nation shall rise up against nation" (MARK 13:8) became "a pair of snowshoes shall rise up against a pair of snowshoes."

oukela," without any regard for the actual meaning of the sounds in Chinese. In different dialects, the phrase "Kekoukela" translates as "bite the wax tadpole" or "female horse stuffed with wax."

- Only one percent of all snowflakes are symmetrical.
- In 1611, German mathematician and astronomer Johannes Kepler wrote a pamphlet entitled *On the Six-Cornered Snowflake*, but was unable to explain the hexagonal shape of snowflakes.
- The world record for most snowfall in twenty-four hours was made on February 7, 1963 when seventy-eight inches of snow fell at Mile 47 Camp, Cooper River Division 4, Alaska.
- The Allied Roofing and Siding Company of Grand Rapids, Michigan, cleaned snow from roofs to prevent them from collapsing from the weight of the snow. In 1979, the company's roof collapsed from the weight of snow on the roof.
- A version of the Bible in an Eskimo dialect included one misplaced let-

X-Ray Glasses

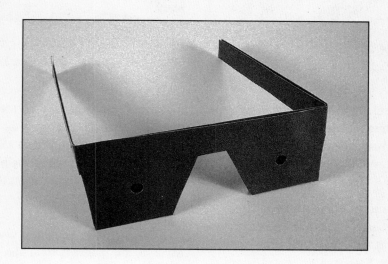

WHAT YOU NEED

- ☐ Two strips of red posterboard (2 inches wide by 17 inches long)
- ☐ Ruler
- ☐ Pencil
- ☐ Scissors
- ☐ Hole puncher
- ☐ Black feather
- ☐ Elmer's Glue-All
- ☐ Two hardcover books
- ☐ Bright light

WHAT TO DO

Fold the first strip of red posterboard in half so the seventeen-inch length is now 8.5 inches long. Place the folded strip on a tabletop with the fold facing to your right. From the top edge of the posterboard strip, use the ruler to measure down $5/8$ inch and draw a line across the width of the posterboard strip. From the right-hand fold, measure in $7/8$ inch and draw a vertical line across the height of the posterboard strip. From the bottom of this line, draw a diagonal line to the top right-hand corner. From the right-hand fold,

 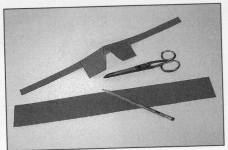

measure in 2½ inches and draw a vertical line across the height of the posterboard strip.

Using the scissors, cut out the large rectangle in the bottom left-hand corner, cutting through both layers of posterboard. Cut out the quadrangle in the bottom right-hand corner, cutting through both layers of posterboard.

Open the pair of glasses and use it as a template on the second strip of red posterboard to create an identical pair of glasses.

Hold the two pairs of red-posterboard glasses together so they line up perfectly, and using the hole puncher, punch a hole in the center of each lens.

Cut two pieces from the feather, each one inch long. Glue the first piece of feather to one of the red-posterboard glasses so that the feather covers one of the holes (without getting any glue on the part of the feather that covers the hole). Glue the second piece of feather over the second hole of that same pair of glasses. Glue the second pair of red-posterboard glasses

to the first pair so the feathers are sandwiched between them and the holes in the lenses line up. Place the two hardcover books on top of the glasses to press the two pieces together until the glue dries.

When the glue dries, fold back the arms of the glasses, wear the X-Ray glasses, and under a bright light, look at your hand through the glasses.

WHAT HAPPENS
You see what appear to be the bones in your fingers.

HOW IT WORKS
The eye sees different wavelengths of light as different colors. Light travels in waves, which usually travel in a straight line. However, when light waves pass through a slit, they diffract (spread out) into curving waves. When light waves pass through several narrow slits (like the numerous spaces created by the complex branching pattern of the feather), they interfere with each other. Where the crest (peak) of one wave meets the crest of another

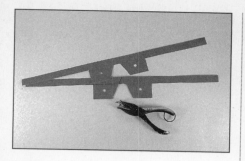

wave—or where the trough (low point) of one wave meets the trough of another wave—the two waves combine and form a bright spot of light. Where a crest meets a trough, the two waves cancel each other out, leaving a dark spot.

Some of the light which would normally be stopped by the edges of your fingers gets bent and reaches your eye, making the edges of your fingers appear semi-transparent. Meanwhile, light from the center of the fingers is not bent enough to reach your eyes, so the center of your fingers appear solid, resembling the bones of your fingers.

BIZARRE FACTS

- X-Ray Specs were commonly advertised in the back of comic books by the Johnson-Smith Company.
- Birds shed their feathers and grow a new set at least once a year.
- Feathers are made of *keratin*, a substance that is also found in the hair of mammals and the scales of fish and reptiles.

- In 1666, English scientist Sir Isaac Newton theorized that light consists of tiny particles called corpuscles that travel in straight lines through space. In 1801, English physicist Thomas Young proved Newton wrong, demonstrating that particles of light travel in waves.
- The 1963 Roger Corman horror film *X—The Man with X-Ray Eyes* stars Ray Milland as a scientist named Xavier who gives himself X-Ray vision and goes insane. The movie co-stars Don Rickles.

DUMB LUCK

In 1895, German physics professor William Konrad Roentgen discovered X rays by accident when he noticed that cathode rays caused a sheet of paper coated with barium platinocyanide to glow—even when the sheet of paper was taken into the next room. Roentgen called the rays X rays because of their mysterious nature.

Yogurt Factory

WHAT YOU NEED

- ☐ 1 quart of half-and-half
- ☐ Glass mixing bowl
- ☐ Cooking thermometer
- ☐ Spoon
- ☐ 3 tablespoons of Dannon Nonfat plain yogurt
- ☐ Clean, empty, glass 1-quart mayonnaise jar with screw-on lid
- ☐ Insulated picnic cooler

WHAT TO DO

Pour the half-and-half into the glass mixing bowl.

With adult supervision and using a cooking thermometer, heat the half-and-half in a microwave oven at fifty percent power for one minute at a time until the temperature of the half-and-half reaches 180° Fahrenheit. (Use the spoon to stir the liquid between time breaks to prevent scalding and skim off any film from the surface.) Remove from the microwave and let cool to about 115° Fahrenheit.

Add three tablespoons of yogurt, mix well, pour into the jar and seal the lid tightly. Place the warm jar inside the insulated picnic cooler, close the lid, and let sit undisturbed for eight hours. Refrigerate when ready.

WHAT HAPPENS
You've made plain yogurt. If you wish, add vanilla extract, or strawberries, peaches, or raspberries to taste.

HOW IT WORKS
Heating the half-and-half kills any bacteria that might otherwise compete with the yogurt cultures. Yogurt contains living bacteria called *Lactobacillus acidophilus*, which multiply exponentially in warm milk. Three tablespoons of yogurt from this batch can be used to start a new batch, ideally within five days.

BIZARRE FACTS
■ Adding the three tablespoons of yogurt to the milk when it is above 155° Fahrenheit may kill the yogurt cultures, which would then prevent the yogurt from forming.

■ Adding more than three tablespoons of yogurt to the warm milk will cause overcrowded bacillus, resulting in a sour, watery yogurt.

CRYING OVER SPILLED MILK

For their satirical operetta *H.M.S. Pinafore* (1878), Gilbert and Sullivan created the pithy lyric, "Things are seldom what they seem; skim milk masquerades as cream."

Bibliography

The Book of Lists by David Wallechinsky, Irving Wallace, and Amy Wallace (New York: Bantam, 1977)

Dictionary of Trade Name Origins by Adrian Room (London, Routledge & Kegan Paul, 1982)

Duct Tape Book Two—Real Stories by Jim and Tim (Duluth, Minnesota: Pfeifer-Hamilton, 1995)

Einstein's Science Parties by Shar Levine and Allison Grafton (New York: John Wiley and Sons, 1994)

Famous American Trademarks by Arnold B. Barach (Washington, D.C.: Public Affairs Press, 1971)

The Guinness Book of Records, edited by Peter Matthews (New York: Bantam, 1993)

How to Spit Nickels by Jack Mingo (New York: Contemporary Books, 1993).

Janice VanCleave's 200 Gooey, Slippery, Slimy, Weird & Fun Experiments by Janice VanCleave (New York: John Wiley & Sons, 1993)

The Joy of Cooking by Irma S. Rombauer and Marion Rombauer Becker (New York: Bobbs-Merrill, 1975)

Jr. Boom Academy by B. K. Hixson and M. S. Kralik (Salt Lake City, Utah: Wild goose, 1992)

Martin Gardner's Science Tricks by Martin Gardner (New York: Sterling, 1998)

100 Make-It-Yourself Science Fair Projects by Glen Vecchione (New York: Sterling, 1995)

PADI Open Water Diver Manual (Rancho Santa Margarita, California: International PADI, 1999)

Panati's Extraordinary Origins of Everyday Things by Charles Panati (New York: Harper & Row, 1987)

Reader's Digest Book of Facts, edited by Edmund H. Harvey, Jr. (Pleasantville, New York: Reader's Digest, 1987)

Science Fair Survival Techniques (Salt Lake City, Utah: Wild goose, 1997)

Science for Fun Experiments by Gary Gibson (Brookfield, Connecticut: Copper Beech Books, 1996)

Science Wizardry for Kids by Margaret Kenda and Phyllis S. Williams (Hauppauge, N.Y.: Barron's, 1992)

Shout!—The Beatles in Their Generation by Philip Norman (New York: Warner Brothers, 1981)

365 Simple Science Experiments by E. Richard Churchill, Louis V. Loeschnig, and Muriel Mandell (New York: Black Dog & Leventhal, 1997)

333 Science Tricks & Experiments by Robert J. Brown (Blue Ridge Summit, Pennsylvania: TAB Books, 1984)

The Ultimate Duct Tape Book by Jim and Tim (Duluth, Minnesota: Pfeifer-Hamilton, 1998)

200 Illustrated Science Experiments for Children by Robert J. Brown (Blue Ridge Summit, Pennsylvania: TAB Books, 1987)

Why Did They Name It . . . ? by Hannah Campbell (New York: Fleet, 1964)

Acknowledgments

I am deeply grateful to my editor, Jennifer Repo, for her astounding professionalism and her passion for this book. I am also indebted to John Duff, copyeditor Cari Luna, and amazing cover artist Brad Weinman.

Once again, a very special thanks to my agent Jeremy Solomon at First Books for his derring-do.

I am also grateful to Matt Strauss, Haleigh Safran, and Sue Solomon at *The View*, Howard Gershen for the paper cup boiler, Zen poet Jeremy Wolff for perfecting the hydrogen balloon experiment, Jim Parish for suggesting the idea in the first place, Girl Scout Troop 1380 for helping me perfect some of the experiments, and Katie Loggia, Emily Graham, Alexa Cohen, Rebecca Leon, and Jay Bruder for their able assistance.

Above all, all my love to my wife, Debbie, and my daughters, Ashley and Julia, for eagerly assisting me with all of the experiments in this book in our garage and for wisely refusing to help me clean up the resulting mess.

About the Author

Joey Green got Barbara Walters to make Green Slime on *The View*, Jay Leno to shave with peanut butter on *The Tonight Show*, Rosie O'Donnell to mousse her hair with Jell-O on *The Rosie O'Donnell Show*, and had Katie Couric drop her diamond engagement ring in a glass of Efferdent on *Today*. He has been seen polishing furniture with Spam on *Dateline NBC*, cleaning a toilet with Coca-Cola in the pages of the *New York Times*, and washing his hair with Reddi-wip in *People*. Green, a former contributing editor to *National Lampoon* and a former advertising copywriter at J. Walter Thompson, is the author of more than twenty books, including *Polish Your Furniture with Panty Hose*, *Clean Your Clothes with Cheez Whiz*, *The Zen of Oz*, and *The Road to Success Is Paved with Failure*—to name just a few. A native of Miami, Florida, and a graduate of Cornell University, he wrote television commercials for Burger King and Walt Disney World, and won a Clio Award for a print ad he created for Eastman Kodak. He backpacked around the world for two years on his honeymoon, and lives in Los Angeles with his wife, Debbie, and their two daughters, Ashley and Julia.

Visit Joey Green on the Internet at www.wackyuses.com